THE GEOGRAPHY OF
Change In South Africa

THE GEOGRAPHY OF
Change in
South Africa

Edited by
ANTHONY LEMON
Mansfield College, Oxford, UK

JOHN WILEY & SONS
Chichester · New York · Brisbane · Toronto · Singapore

Other Wiley Editorial Offices

John Wiley & Sons, Inc., 605 Third Avenue,
New York, NY 10158-0012, USA

Jacaranda Wiley Ltd, 33 Park Road, Milton,
Queensland 4064, Australia

John Wiley & Sons (Canada) Ltd, 22 Worcester Road,
Rexdale, Ontario M9W 1L1, Canada

John Wiley & Sons (SEA) Pte Ltd, 37 Jalan Pemimpin #05-04,
Block B, Union Industrial Building, Singapore 2057

ISBN 0-471-95619-8

Typeset in 10/12pt Sabon by Acorn Bookwork, Salisbury, Wiltshire
Printed and bound in Great Britain by Bookcraft (Bath), Avon

Contents

Part III   SOUTH AFRICA IN SOUTHERN AFRICA

Figures

Tables

About the Contributors

Tony Buckle was formerly a postgraduate student in the Department of Geography at Queen Mary and Westfield College, London University, where he completed an M.Phil. thesis on land–power relations in the Eastern Cape region of South Africa. He is now working for an M.B.A. in Barcelona.

Anthony J. Christopher is Professor and Head of the Department of Geography at the University of Port Elizabeth. He has published seven books including *Southern Africa* (Dawson, 1976), *South Africa* (Longman, 1982), *Colonial Africa* (Croom Helm, 1984), *The British Empire at its Zenith* (Croom Helm, 1988), and *The Atlas of Apartheid* (Routledge, 1994). His current research concerns the experience of indigenous peoples in Commonwealth countries of European settlement.

James Drummond is a Lecturer in Geography at the University of the North West. His research interests include rural and agricultural development and the historical geography of southern Africa.

Roddy Fox is a Lecturer at Rhodes University, Grahamstown, South Africa. He has published on the urban and settlement geography of East Africa, and more recently on urban and electoral issues in South Africa.

Richard Gibb is a Senior Lecturer at Plymouth University. He did his doctorate at Oxford University on the 'Anglo-French frontier region', and has since published widely on the Channel Tunnel and other European issues. A Visiting Lectureship at Cape Town University, and an ESRC research award, enabled him to develop his interests in regional economy in the southern African context.

Anthony Lemon is a Lecturer in Geography at Oxford University, where he is Fellow and Senior Tutor of Mansfield College. He has held visiting lectureships and research fellowships at several southern African universities. His earlier books include *Apartheid: a Geography of Separation* (Saxon House, 1976), *Apartheid in Transition* (Gower, 1987) and *Homes Apart: South Africa's Segregated Cities* (Paul Chapman, 1991), and he is co-editor of *Studies in Overseas Settlement and Population* (Longman, 1980). He has also published many articles on the urban, social and political geography of apartheid, and has recently made a special study of educational change in the new South Africa.

Malcolm Lupton is a Lecturer in Geography at the University of the Witwatersrand, Johannesburg. His research interests are in urban studies, especially housing policy in the coloured townships of the Johannesburg region, as well as urban and social theory and Anglo-American geographical thought.

Stuart Murphy is a Senior Research Analyst, Premises Division, Standard Bank of South Africa. He has also worked for the South African Institute of Race Relations, where he concentrated on demography, population and urban policy. His current research interests concern the geography of marketing and finance in South Africa.

Garrett Nagle is Head of the Department of Geography at St. Edward's School, Oxford. He completed his doctoral thesis at Hertford College, Oxford on malnutrition in the Zwelitsha area of Ciskei. He has paid several visits to the region, conducting field research into health and health care issues.

Edward Ramsamy is currently completing a Ph.D. in urban and regional planning at Rutgers University in the USA. His primary research interests concern the implications of changes in the political economy of international relations for planning theory and practice. He is also editor of *Common Purpose*, a Rutgers University journal which aims to encourage dialogue on cultural pluralism.

Christian M. Rogerson is Professor of Human Geography and Head of the Department of Geography at the University of the Witwatersrand, Johannesburg. Recent co-edited works include *South Africa's Informal Economy* (Oxford University Press, Cape Town, 1991), *Geography in a Changing South Africa: progress and prospects* (Oxford University Press, Cape Town, 1992) and *Finance, Institutions and Industrial Change: Spatial Perspectives* (Walter de Gruyter, Berlin, 1993). His current research focuses on local development, industrial restructuring and the informal economy in South Africa.

David M. Smith is Professor of Geography at Queen Mary and Westfield College, University of London. He has published extensively on geographical aspects of inequality, human welfare and social justice. His most recent book, *Geography and Social Justice*, was published by Basil Blackwell in 1994. His work on South Africa includes *Apartheid in South Africa* (Cambridge University Press) and *The Apartheid City and Beyond* (Routledge). He lived in South Africa in 1972–3, and has spent a number of subsequent visits undertaking research and lecturing in various South African universities.

Coleen Vogel is a Lecturer in the Department of Geography and Environmental Studies at the University of the Witwatersrand, Johannesburg. She is completing a doctorate on the socio-economic consequences of drought in southern Africa. Her research interests include rural and environmental problems associated with development, environmental degradation and climatology.

Introduction

ANTHONY LEMON

GEOGRAPHY AND APARTHEID

For geographers, apartheid has long exercised a peculiar fascination. South Africa is in some senses a microcosm of the processes of development, under-development and exploitation which have characterized global economic rela-tions. It exemplifies First World–Third World relationships in an extreme form, with pressures of shared space leading the white minority to erect a complex, rigid system to maintain control and protect its economic privileges: privileges which are much more easily and less conspicuously protected under the structures of long established nation states (and more recent trading blocs, including the European Community) comfortably distant from the Third World. More perhaps than any other political system, apartheid in South Africa has created its own 'applied geography'—economic, social and political—at micro-, meso- and macro-scales. Tragically misguided in the eyes of most of humanity, apartheid was pursued with extraordinary zeal and even a distorted idealism by Afrikaner nationalists from 1948 onwards (de Klerk, 1975, pp. 241–2).

Few objects of human geographical study have been left untouched by this all-pervading system (Lemon, 1976; Smith, 1990; Christopher, 1994). It has sought to control that most fundamental geographical expression, the distribution of population, most notably by means of racial land allocation and the restriction of rural–urban migration by black South Africans. Less strenuously, it has attempted to modify the distribution of economic activity, and especially manufacturing employment, to provide some support for its desired distribution of population. This led to the building of new towns as 'homeland' capitals— some like Mmabatho now large enough to attract their own migration flows— and as industrial growth points, often in unlikely locations such as the Qwaqwa capital of Phuthaditjhaba, now a settlement of over 200 000 people at the end of a tarred road under the shadow of the Drakensberg escarpment (Pickles and Woods, 1992).

In the cities where even apartheid conceded that the various elements of a plural society must come together in the market place (Furnivall, 1948), it sought to keep them apart socially by fundamentally restructuring the residential geo-graphy in the 'group areas' of the apartheid city (Lemon, 1991; Smith, 1991). Even at the micro-scale, 'petty apartheid' sought to minimize or eliminate the most casual of contacts at what was deemed the most sensitive interface, by providing separate (but not equal) facilities for whites and 'non-whites' for everything from toilets to park benches, buses to post office counters, beaches to station platforms.

The power which undergirded this system sought to perpetuate itself by creating a new political geography. Group areas were both an expression and a reinforcement of power relations, as John Western (1981) showed in his study of Cape Town's 'outcast' coloured people. In the 1980s they were to become the foundation of a racially ordered local government system (Heymans and Tote-meyer, 1988), albeit one which never gained legitimacy among blacks and remained largely inoperative among coloured people and Indians. At the macro-scale, 'grand apartheid' sought to solve the 'problem' of three-quarters of the country's population by creating black 'homelands', enlarging and consolidating 19th century reserves for the purpose. These territories in the periphery of the South African space-economy were presented as potential independent states in which all blacks would enjoy citizenship rights: this was, according to Pretoria, simply the same process of decolonization which was being pursued all over Africa (see Chapter 1).

For coloured people and Indians, both largely urbanized, a territorial solution remained elusive. An ambivalent policy of 'parallelism' was pursued in the 1970s, but it was a South African Prime Minister, B.J. Vorster, who pointed out the impossibility of having two sovereign governments in the same geographical territory. One of his successors, P.W. Botha, tried to resolve the problem with what purported to be a power-sharing constitution in 1984, embodying a division between 'general affairs' (with in-built structures ensuring white domination) and 'own affairs' for white, coloured and Indian houses of a tricameral parliament. Group areas provided the territorial basis for the administration of 'own affairs', while ultimate white control was secured by the inclusion of budgets for these parliaments as a 'general affair'. Even this nudge in the direction of power-sharing proved too much for right-wing whites who broke away from the ruling National Party to form the Conservative Party in 1982.

REFORMING APARTHEID

The 1984 constitution heralded what has been widely labelled an era of 'reform', typifying (and possibly influenced by) a Harvard professor's model of 'Reform and stability in a modernizing, multi-ethnic society' (Huntington, 1981; Lemon, 1987a, pp. 326–8). In the mid-1980s the new State President, P.W. Botha, displayed both vision and myopia. He recognized the impossibility of going on as before but believed apartheid capable of reform to make it sufficiently acceptable, internally and externally, to continue without disturbing the fundamentals, including ultimate white control. His prime weapon in this flawed vision was not political but economic.

It was not political because it could not afford to be. To add a fourth chamber to the tricameral parliament, however circumscribed the definition of the non-homeland black population, would seriously compromise and eventually undermine the structures of white power. To move towards a unified system of local government would have had the same effect in the cities. To accept demands for a unitary education system would inevitably have undermined the standards to which whites were accustomed. To scrap the Group Areas Act, described by P.W. Botha as his 'bottom line', would have been to remove the keystone from the complex edifice of apartheid, threatening to bring down much more with it.

Botha's reforms, while not insignificant, made little concession to such populist demands. Typically, each reform recognized a portion of geographical reality and tried to isolate it from the wider whole. Thus the abolition of racially discriminatory influx control measures in 1986 recognized the inevitability, even the economic desirability, of black urbanization, but left group areas and racial allocation of housing land in cities in place, while attempting to secure the 'orderly' channelling of black urbanization to settlements of the state's own choosing (Mabin, 1989). At local level, Regional Services Councils brought representatives of different race groups together but as representatives of the racially constituted organs of local government and with voting proportionate to consumer power, not population (Lemon, 1992).

In effect, Botha sought to buy legitimacy, or at least acquiescence. Thus Regional Services Councils were above all about redistribution—improving infrastructure 'where the need is greatest', using revenues provided by new taxes on businesses (use of property rates was deemed too sensitive for the white electorate). In housing, despite an appearance of retreat by the state (again with the white electorate in mind?) in terms of shifting burdens of provision to the private sector, the actual contribution of the state to black housing increased substantially. Most important of all, the state massively increased its expenditure on black education, beginning the huge task of closing the racial expenditure gap (see Chapter 7).

In these and other ways, it was hoped that evidence of tangible benefits from the reform process would induce significant numbers of urban blacks at least to acquiesce in the system. Such an attempt was doomed to failure, first and foremost because it ignored the essentially political nature of black demands for full political freedom, equality and citizenship. Blacks could not be so easily induced to sell their birthright for a mess of potage. Secondly, the resources generated by the South African economy were not remotely sufficient for the purpose in hand. Relatively small but tangible improvements in their lives may indeed encourage most blacks to accept, even grudgingly to support, a black or black-led government in the 'new South Africa', although the ability of the new government to deliver sufficiently in this respect remains in doubt, as several contributors to this volume show. What should have been clear in the late 1980s, however, was that such benefits stemming from the 'reformed apartheid' of a *white* government could only be perceived as 'crumbs from the rich man's table' and as such could never be the basis of popular legitimacy or even acquiescence.

Botha made clear the limits of his vision in his much heralded 'Rubicon' speech to the National Party congress in Durban in 1985. Far from signalling his intention of crossing the Rubicon, as the media build-up led international opinion to expect, he warned 'don't push me too far', and forecast that the results of majority rule in South Africa would be 'faction fighting and chaos'. The speech led directly to the refusal of American and European banks to roll over short-term loans to South Africa, which, after renegotiation, led to net capital outflows during the late 1980s, as loans were repaid and no new ones secured. Thus capital inflows, arguably the lifeblood of a semi-developed economy such as South Africa's (as exemplified by the rapid growth of the 1960s), were reversed, with little prospect of resumption without fundamental political change. The economic stagnation of the early 1980s was accentuated, and the capacity of the economy to employ its labour force steadily eroded (see Chapter 9). Some leading capitalists even began to think the unthinkable—could they be better off under the ANC?—and went to meet ANC leaders to find out, to the fury of P.W. Botha, who (rightly) perceived their actions as undermining his government.

DISMANTLING APARTHEID AND NEGOTIATING A NEW DISPENSATION

Such was the political and economic background which greeted F.W. de Klerk on his inauguration as State President in 1989. It seems likely that economic prospects dominated his calculations in preparing the announcement of fundamental changes in February 1990: the unbanning of the African National Congress (ANC), the Pan Africanist Congress (PAC) and the South African Communist Party, the dismantling of apartheid and the negotiation of a new constitutional dispensation. It was clear that he had no intention of simply surrendering white power and that a long, hard period of negotiation was in prospect, between government forces which still had economic, political and military power and popular forces which had both internal legitimacy and external support which could continue to damage the South African economy.

Many hurdles had to be surmounted before negotiations could even begin, including questions concerning the release of political prisoners, the return of political exiles and the suspension of the armed struggle. The ANC and PAC had to establish themselves as domestic political organizations. The structures of the negotiating forum, and the question of who should participate, had to be agreed. With such a formidable agenda, it should be no surprise that the whole process took four years, although this inevitably angered blacks in the townships for whom the basic issues were all too clear.

Arguments over the nature and causes of political violence repeatedly delayed both the start and the progress of negotiations, most notably when the Convention for a Democratic South Africa, the first negotiating forum inaugurated in December 1991, was effectively terminated by the withdrawal of the ANC in June 1992 amid accusations (which subsequently proved well founded) of government complicity in violence against its supporters. The government and the ANC committed themselves to further talks in September 1992, but these remained effectively bilateral for several months, to the fury of Chief Buthelezi, Chief Minister of KwaZulu and leader of the Inkatha Freedom Party. Multi-party talks were not reconvened until April 1993, in the shape of the 26-party Multi-Party Negotiating Council (MPNC).

This left all the big decisions to be made in a matter of months. These included the nature of decision-making in the interim government, the procedures for drawing up and approving a final constitution, the division of powers and constitutional relationship between central government and regional authorities, the number and boundaries of the regions, the nature of local government and the details of the electoral system. The result is, ironically given the length of the lead-in period to negotiations, a constitution which bears the hallmarks of rushed decision-making and last-minute compromise. Even the regional boundaries are in some cases provisional, and may be altered by referendum (see Chapter 2), while the impossibility of compiling an accurate electoral register in the time available dictated the far from ideal use of a list system at both national and regional levels.

Such reservations should not detract from the remarkable achievement represented by a constitutional agreement negotiated by representatives of the vast majority of black, white, coloured and Indian South Africans. The parties concerned sat down together in a country with no tradition at all of negotiation between black and white. They did so without the benefit of any foreign broker or facilitator, yet they managed to agree not only on broad principles, but on much finer detail. A culture of negotiation has been created on the most unpromising of foundations. There are few if any historical or contemporary precedents for such an achievement: certainly not in Northern Ireland or the former Yugoslavia, and even the much-hailed 1994 breakthrough between Israel and the PLO represents only a fraction of the progress made in South Africa.

CHALLENGES FACING THE NEW SOUTH AFRICA

The contributions in this volume address three sets of challenges facing the new South Africa. The first concerns the immediate challenges of constitutional transition. A new constitution was agreed by most parties in the MPNC in November 1993, but it is only an interim constitution, to be superseded within five years by a permanent constitution. The first two chapters of this book address key issues of political geography which have emerged during the constitutional negotiations, namely the nature of the nation-state and the shape and powers of regions in the new South Africa. These will almost certainly remain important matters of debate for the new constitution-making body, consisting of the National Assembly and Senate sitting together, elected in April 1994.

The bulk of the book is devoted to the challenges of domestic policy faced by a new government, which is confronted by the exaggerated expectations of its electorate (expectations which the ANC did little to dispel in its election campaign) in the context of a stagnant economy, widespread poverty and unemployment, extreme inequalities and limited resources. Three and a half centuries of white rule have allowed a minority of the population to acquire living standards far beyond those which the country can afford for the majority of its population. Yet these standards have inevitably become the focus of black aspirations and, in the case of the important and growing black middle class, of their immediate expectations.

The third and final section of the book is concerned with the challenge of reformulating economic and political relations with South Africa's regional neighbours. Hitherto co-operation of these ideologically diverse states has been promoted largely by their common opposition to apartheid and their wish to reduce dependence on South Africa, although Swaziland and Malawi have been somewhat ambivalent on the latter (Lemon, 1987b). With the retreat of Marxism–Leninism in Mozambique, Angola and (in so far as it was ever practised there) Zimbabwe, there has been a degree of ideological convergence in the region which may ease the process of co-operation. But with the ending of apartheid in South Africa, a major part of the *raison d'être* for such co-operation has disappeared. As the new South Africa takes its place in the Southern African Development Community (SADC), new tensions are likely to surface. Economic dominance by a country with three times the GNP of the rest of SADC put together will still arouse resentment. A country which is so much stronger economically than its neighbours will inevitably be expected to assist them, yet its own problems and the expectations of its people will almost certainly preclude philanthropy.

CONSTITUTIONAL ISSUES

Noting the enduring centrality of the nation-state concept in political geography, Christopher stresses the fragility of multi-national states in the present century, suggesting that state disintegration rather than integration has become the norm. He considers the apartheid model of partition in South Africa based on the pattern of landholdings, its limited correspondence with the country's language geography despite the enforced resettlement of population both from 'white' areas and between homelands, and the massively unequal distribution of resources between the homelands and 'white' South Africa. The homeland partition failed to gain either external support or internal legitimacy, and by the late 1980s the National Party was reconciled to its failure, although it failed to present a clear alternative.

Despite this unpromising history of partition, and the equally unrealistic partition proposals advanced by various white right-wing groups, Christopher suggests that there may be a place for partition in the politics of post-apartheid South Africa, stressing that he is offering 'an academic speculative warning as to the potential for fragmentation, not as a political programme'. Deviation from Africa's colonial boundaries, so long successfully resisted by the OAU, is beginning to occur; in South Africa, the most likely possibility is the self-determination of the largest ethnic group, the Zulus, within a clearly defined colonial boundary (Natal). Recognizing that the Inkatha Freedom Party currently represents only a minority of Zulus, Christopher is focusing his comments on the longer-term possibility that 'South Africa might be able to achieve a more prosperous and peaceful future without the destabilizing need to coerce a potentially hostile and disruptive population in Natal'.

In the current political climate, such a view will be highly controversial, whatever the time perspective. If, as increasingly appears to be the case, Buthelezi is manipulating his Zulu supporters rather than merely mobilizing a Zulu nationalism genuinely anxious for political autonomy, then current events lend only limited support to Christopher's speculation. But the potential disruptive power of Zulu ethnonationalism is undoubted, and, along with ethnicity in South Africa generally, it could easily come to the fore as a potent political force in the face of widespread dissatisfaction with a future government, now that the common enemy of apartheid has been removed.

Ethnicity is unquestionably far from a spent force in South Africa, and in Chapter 2 Fox's study of the regional constitutional debate illustrates its potential significance. He traces the series of proposals emanating from the ANC and the National Party in the period 1991–3, including their eventual submissions to the Commission for Demarcation/Delimitation of States/Provinces/Regions. The very title of this body indicates the provisional context in which it was working, delimiting regions without knowing what their powers were to be or what arrangements if any were to operate for the sharing of revenue. Fox examines the stances of the ANC and the National Party (NP) on the question of regional powers and explores in some detail the criteria used in formulating the various regional proposals, including those of the Commission itself in its two reports. He points to the significance of electoral considerations for both the ANC and the NP and to the distribution of wealth between regions for the ANC in particular. He also draws attention to the various contentious areas where provision has been made for post-election referendums, if desired, and possible boundary changes; such is the uncertainty that these could even lead to the amalgamation of two regions (Northern Cape and Western Cape) and the division of another (Eastern Cape).

The interim constitution which was finally agreed represents a compromise between the centralism of the ANC and the federalism of the NP. The latter's support for federalism is somewhat ironic given its own record of increasing centralization—taking over the administration of black townships from local authorities in 1972, for instance, and abolishing elected provincial councils in 1985. Faced with the prospect of losing power at the centre, however, the NP clearly wished to limit the powers of its successor and hoped to gain control of one or more regional governments—the Western Cape, with its white/coloured majority appeared to be its best chance. The ANC not unreasonably argued that the formidable task of structural transformation facing a post-apartheid government demands strong central powers; certainly this accords with what appear to be its statist instincts.

The constitutional compromise is that of 'regionalism'—not a clearly definable term in the political geographer's lexicon. In many respects it is thoroughly federal: the regions have extensive powers (see Chapter 2), they have their own elected governments with a premier and executive council and even the power, within the limits of nationally agreed principles, to write their own constitutions.

The key qualifications are those which allow the national parliament to override provincial laws in specified circumstances (Republic of South Africa, 1993, p. 86), and the ambivalence of the provision for revenue-sharing (p. 100). The latter merely states that 'a province shall be entitled to an equitable share of revenue collected nationally to enable it to provide services and to exercise and perform its powers and functions' (section 155, subsection 1); the nature of the revenue-sharing formula remains to be decided, and the presumed power of central government to change it is an implicit limitation on regional power.

Provisions for central government to override the provinces are capable of very broad interpretation, especially perhaps the reference to 'the maintenance of national security' (section 26, subsection 3d). It may well be that the central government will seldom use these powers, in which case the new South Africa will be a *de facto* federation; if it has frequent recourse to them, however, the conflict potential is obvious in a state which will have all the trappings of federation.

DOMESTIC POLICY ISSUES IN THE NEW SOUTH AFRICA

It is appropriate that this section of the book begins (Chapter 3) with Smith's discussion of redistribution and social justice, for these are the concerns which underpin the chapters on land, health, education, housing and employment which follow, as well as the ethical questions raised by environmental issues. Recognizing the difficulty of saying what kind of post-apartheid society might most persuasively be defined on grounds of social justice, Smith explores relevant theoretical deliberations, arguing that a form of egalitarianism provides an appropriate universal framework to guide thinking about social justice in South Africa but that its application requires attention to the particular inheritance of apartheid.

Smith argues that the pace of equalization must be greatly accelerated and explores three possible redistributional strategies: reallocation of public expenditure, redistribution of wealth and structural change. He concludes that the prospects for rapid equalization are not encouraging; constraints on redistribution are severe, and there is a danger that transfers of resources may not flow right down the socio-economic scale and may be geographically selective, favouring residents of formal urban areas in metropolitan regions.

Smith's analysis reveals a yawning gap between what is implied by theoretical considerations of social justice and what is most likely to happen, which is that South Africa will steadily come to resemble more closely a normal capitalist society, its inherited inequalities interpenetrated by class but with an unusually large 'underclass'. While this scenario will not debar South Africa's return to the 'community of nations', sustainable claims to social justice would require redistribution to take place at a scale and pace of change, including elements of structural change, which is scarcely conceivable in terms of recognized constraints and currently perceived political realities.

Land issues pose some of the most complex challenges facing the new government. Just how complex is revealed by Buckle's detailed study (Chapter 4) of two former mission communities, Mgwali and Lesseyton in the Border Corridor of the Eastern Cape. His study must be viewed in the broader context of the 'land question', which embraces questions of restitution of land to those forcibly dispossessed and resettled, redistribution of much of the 70 per cent of land still under white ownership and reform of land tenure, especially in the homelands. The de Klerk government's approach to restitution claims falls far short of the demands of social justice, but the practical problems of dealing fairly with land claims will be immense. Any cut-off date for consideration of claims will be arbitrary, and the earlier the date, the more complex will be the problems.

The NP government repealed the Land Acts of 1913 and 1936 in 1991, thus removing restrictions on black purchase of 'white' land, but little *de facto* land redistribution is likely to occur through purely market mechanisms. Acquisition of white land poses questions of the nature and extent of compensation and the basis of selection. Abandoned and underutilized farms are obvious candidates. Leased land, where the owner is not working his own land, might be expropriated, or corporately owned farms broken up, although the latter are often simply a convenient structure within which individual farmers conduct their business. Farms with income below a certain figure might be expropriated, assuming them to be either unproductive or dependent on other sources of income. But all these suggestions ignore the geographical distribution of demand, which is greatest in the areas adjoining the homelands. Farms which are broken up into many smaller units would also need higher levels of infrastructure and servicing if they are to be fully productive.

Equally problematic will be the selection of recipients, a delicate problem which is potentially critical to the productivity of redistributed land: identifying those who will actually use land productively will not be easy. Models of settlement and the nature of land tenure also pose complex questions.

Land tenure is also an issue in the homelands. It has often been argued that reform of the existing traditional system of landholding, replacing it with leasehold or freehold, is an important component of any rural development strategy. However, such views appear to be based on a misunderstanding of the traditional system, based on the label 'communal'; in practice it provides individuals with a high degree of security of tenure. Low productivity levels, and indeed the absence of farming at all in many cases, are explicable in terms of the unaffordable cost of inputs, the inadequacy of rural transport and supply infrastructures, the lack of agricultural and management skills (in part a consequence of migrant labour) and the higher returns from wage labour. In the short term, state resources are probably better devoted to raising agricultural productivity in the homelands than to reform of traditional tenure which would probably increase landlessness at a time of high unemployment.

Buckle goes beyond questions of production and national land divisions between black and white. He shows land–power relations to be informed by both the ethics of the colonial private property regime and the traditional African social land ethic. In both the communities he studied, a distinctly land-based social hierarchy has developed. This is seemingly upheld by all, with the effect that the landless majority are effectively contributing to their own marginalization: while seeking changes in their individual positions in the land–power structure, the system itself was not challenged. In such a situation, Buckle argues, community demands and social justice prove irreconcilable. The fact that land is much more than an economic commodity has significant implications for land redistribution in South Africa: the return of land is seen as a necessary part of the return of political power. But it is more than a question of transferring land from whites to blacks, important though that process is. Local divisions between black and black are equally important, yet extraordinarily complex and intractable.

Land degradation and soil erosion have long affected parts of South Africa, especially the homelands, but in recent years several other environmental issues have come to the fore. In Chapter 5 Vogel and Drummond argue that 'green' issues of land and conservation cannot be separated from urban 'brown' environmental problems such as poor quality water, inadequate sanitation, waste removal, air pollution and land degradation which have their origin in the political economy of apartheid. Land degradation in the homelands has been accentuated by the racial allocation of land itself and by forced removals and resettlement. The impacts of drought are accentuated by poor management and lack of development. State subsidies have until recently encouraged whites to farm in marginal areas. Dependence of the poor on wood for fuel leads to deforestation. Many urban environmental problems including air and water pollution are linked to the lack of housing, land and essential services including electricity, water supply, sanitation, sewerage and waste removal.

It is a measure of South Africa's relative development in African terms that these environmental problems are receiving increasing attention from the government and other organizations, including political groups. Vogel and Drummond highlight the need for research, environmental education and alternative approaches to enable local populations either to revive past capacity or to create new ways of managing the environment. Such conclusions raise broader resource questions. Faced with so many pressing needs in the new South Africa, it may be difficult to give priority to environmental issues, despite growing environmental awareness. More crucially, however, it is apparent that so many of the issues discussed relate to wider problems of poverty and inequality in both urban and rural environments and cannot be solved apart from their root causes in terms of land, housing, health and employment issues.

Health has not been widely studied by geographers in South Africa, and in Chapter 6 Nagle draws attention to several problems in the data sources. It is clear, however, that the country's disease profile and mortality rates follow clearly defined racial lines, reproducing First World/Third World patterns, with degenerative diseases predominating among whites and contagious diseases among coloured people and blacks, although degenerative diseases are growing among blacks in larger urban areas. HIV infection is also a growing problem, especially among blacks where transmission is largely by heterosexuals. Hepatitis B infection, although more widespread and potentially serious, has attracted much less attention.

Racial cleavages in health patterns reflect both general socio-economic inequalities and highly unequal access to health services administered within an apartheid framework characterized by urban bias, excessive fragmentation, white dominance and misallocation of funds in favour of 'white' curative hospitals and the private sector rather than the preventative health care demanded by the health profile of blacks. However, Nagle points out that the development of a dichotomous health care system of preventative care for the poor and curative care for the well-off is not in the spirit of the new South Africa. Health, like other sectors, must compete for resources; current high levels of violent conflict also add to health costs, as do growing levels of drug and alcohol abuse, while the cost of AIDS-related expenditure will consume an increasing share of the health budget in the next two or three decades.

Nagle argues that health policies should treat South Africa as a primarily Third World country undergoing rapid urbanization and social transformation, with the majority of diseases and deaths related to poor socio-economic conditions, poor housing, overcrowding and low incomes—a conclusion strikingly similar to that of Vogel and Drummond in their analysis of environmental problems. Given resource constraints, however, Nagle recognizes that health services in the foreseeable future are likely to be strongly influenced by the nature and location of existing facilities: the legacy of apartheid is a straitjacket from which escape will be slow and painful.

There is a similar message in Chapter 7, where Lemon argues that resource availability will act as a fundamental constraint on the transformation of educational opportunity for the poor. The transitional policies of the de Klerk government left education in a phase of concession and compromise, reflected in changing official policies on admission to and financing of white schools. In effect the government moved towards a policy of limited desegregation and semi-privatization, resisting the implications of equal per capita spending under a new dispensation by imposing upon parents the obligation to pay fees in order to maintain the standards to which they have been accustomed. The gradual replacement of racial divisions with class divisions is the inevitable result of such policies.

Black education has been a scene of struggle in urban areas since the 1970s, and expectations of a post-apartheid government are high. Existing inequalities in terms of resources, teacher to pupil and classroom to pupil ratios, teacher qualifications, technical equipment and books remain high, despite considerable progress in closing the racial spending gap. Given the size of the existing education budget, improvements for blacks will depend largely if not entirely upon redistribution within this budget. Both the NP government's Education Renewal Strategy and the reports of the National Education Policy Investigation must be considered in the light of this constraint.

Lemon draws attention to several geographical and other constraints on the potential for redistribution in South Africa's education system, including racial population ratios, differential population growth, the number of blacks currently not in school, problems of teacher supply and mobility, the immobility of school plant and the dismal state of black scientific and technical education. The greatest challenge arises from the massive rural–urban gap: within the apartheid education budget, homeland blacks and those on white farms received far fewer resources than even the township schools which have been the centre of struggle.

In the face of such daunting problems, it is vital (as in other spheres) to develop a national strategy establishing priorities which enjoy a broad basis of consensus. The National Education Forum, finally established in August 1993, will play a key role in this respect. National economic growth will be an essential prerequisite to serious progress towards a socially just education system, and an increased and better directed contribution from the private sector is also essential. Lemon concludes on a cautionary note, emphasizing that education *per se* will do little to undermine the inequalities of capitalism.

Those inequalities are nowhere more starkly reflected than in the contrasts between middle-class, still predominantly white, suburbs and the townships and informal settlements where the vast majority of blacks continue to live. Even the casual observer cannot fail to see both the contrast and the high degree of segregation in South Africa's smaller towns, where the box-like houses of a typically treeless township are clustered tightly together on the veld, often a mile distant from the 'white' town and its commercial facilities. In larger cities, it may take more effort to find the places in which blacks live, but the numbers, and the rate of increase, are much greater than in smaller towns. By 1994 over eight million people lived in informal housing, mainly in metropolitan areas, including nearly three million in the Pretoria–Witwatersrand–Vereeniging (PWV) area, and approaching two million in greater Durban. Many of these people are 'displaced urbanites', living across the homeland borders which have acted as containing walls under the influx control system which operated until 1986. Those with jobs often face long and difficult journeys as 'frontier commuters' (Lemon, 1982). Others, especially since 1986, have filled patches of open ground in townships, or found places to stay in 'white' areas (Crankshaw and Hart, 1990; Crankshaw, 1993). The Urban Foundation (1990) projects growth of 600 000 Africans per annum in metropolitan areas. Two-thirds of this growth will result from natural increase rather than migration which suggests a limited role at best for regional policies designed to reduce pressures on metropolitan areas.

Urban housing is therefore a critical problem for the new South African government, and once again one where 'the shadow of apartheid planning will be evident . . . for decades to come' (Wills, 1991). Lupton and Murphy (Chapter 8) examine the gyrations of black housing policy in relation to the wider geopolitical concerns of apartheid planners, and go on to consider the competing contemporary housing policies and urban strategies being advanced by both 'establishment' and 'alternative' institutions. The former include the World Bank, the Urban Foundation, the Independent Development Trust, the Development Bank of Southern Africa, the South African Housing Trust and the NP government's own proposals emanating from the de Loor Task Group. All support a broadly market-defined approach, although some place more emphasis on housing delivery, while others seek to build up the capacities and skills of the intended beneficiaries. Alternative, more process-focused, approaches aim to build the capacities of communities to enable them to gain a measure of control over the development processes which affect them. Two broad stances are identified here: an anti-statist, neo-populist approach, exemplified by housing credit co-operatives, and a social-democratic, neo-corporatist approach where the focus is on state housing programmes.

Lupton and Murphy are critical of what they see as a widely shared ideological bias against state-provided housing, but they concede that, given the present state of the debate, a future housing policy is most likely to take shape around a broadly defined market-oriented framework, albeit with a greater emphasis on 'process' to complement product delivery. The National Housing Forum, which has brought non-governmental stakeholders into housing policy formulation since its launch in August 1992, is likely to play an important role under the new government.

Unemployment is undoubtedly the most pressing challenge of all for South African policy-makers because its current and growing magnitude adversely affects progress towards social justice in all other sectors. Rogerson (Chapter 9) begins by documenting and explaining the deteriorating job situation in the formal sector, noting that an estimated 53.5 per cent of the potential workforce lacked formal jobs in 1991. He then discusses several policy proposals to ameliorate the mounting crisis, including short-term job creation programmes, especially in construction and public works, and the role of the informal economy. Implementation of the former demands restructuring of public-sector expenditure at the macro-level to favour the unemployed and poor. The informal economy already accounts for perhaps three to five million people and is both complex and heterogeneous in its make-up, varying from informal economies of bare survival to those of growth but with survival enterprises preponderant. For these, welfare-assistance measures are most appropriate. Deregulation in the past decade has helped growth segments of the informal sector. Small-scale manufacturing, construction and transport enterprises are the most promising, and Rogerson suggests a number of possible policies aiming to 'grow' these branches of the informal economy.

Beyond the needs of short-term job creation and managing the diversity of the informal sector, the major task will be to reconstruct the South African economy towards a long-term, sustainable and employment-creating growth path. Rogerson examines the nature of such a development strategy, guided by the principle of 'democratizing economic growth', in terms of the ANC's proposals for agriculture and industry. He offers a number of comments on the regional dimensions of such strategies, identifying promising areas for a viable smallholder agricultural sector and spotlighting key implications of the withering away of the apartheid regional development programme and the accompanying rise of many local development initiatives. The major focus of industrial and other employment growth is likely to be in the leading metropolitan regions.

SOUTH AFRICA IN SOUTHERN AFRICA

The demise of apartheid is the major change to affect the southern African region in the 1990s but not the only one. Increasing ideological convergence since the ending of the Cold War has already been noted; this has been accompanied by progress towards the democratization of political institutions and greater emphasis on economic liberalization. Prospects for international development aid are less favourable than in the mid-1980s: the possibility of a 'Marshall aid' plan for the region once apartheid ended was canvassed then but is no longer on the agenda, given the needs of Eastern Europe and the states of the former Soviet

Union and the ending of superpower conflict which greatly reduces the region's strategic significance. In this changed context, the Southern African Development Coordination Conference (SADCC) reconstituted itself as the Southern African Development Community (SADC) in August 1992, seeking to become a more effective instrument of regional integration.

Ramsamy (Chapter 10) examines economic and political relationships between South Africa and the SADC states in the context of these changes. He first proposes a framework for conceptualizing regional alliances and relations of dependency and interdependency in the region. Challenging the dependency and modernization approaches, Ramsamy adopts a comparative political economy perspective which stresses the importance of the structure, function and policies of the individual states and their domestic political economies. Using this framework, he gives an overview of the structure and objectives of SADCC and critically examines its successes and failures. Partial successes included SADCC's emphasis on transport projects and its attraction of development aid. But it made little progress towards reducing dependence on South Africa, and national planners and the domestic elites they serve have frequently proved unwilling to put regional interests first. The political and bureaucratic weaknesses of many SADCC states, together with regional and/or ideological factionalism, both compounded and assisted South African destabilization.

Moving to consider the complexities of South Africa's future involvement in SADC, Ramsamy notes the tensions which underlie political rhetoric concerning South Africa's future role in the region: fears of potential domination, doubts about the commitment of the ANC to regionally based development initiatives in the face of pressing domestic priorities and the future of migrant labour flows to South Africa. If the growth potential and purchasing power of a regional market are to be realized, both physical infrastructure and regional co-ordination agencies will need further development. To address historical legacies of uneven development, Ramsamy argues the need for an urban and regional development strategy which transcends national boundaries and a regionally dispersed production system. Such developments go far beyond anything achieved by other regional organizations in the developing world, and at a time when changes in South Africa have already led to tensions within SADC, they seem unlikely to materialize in the foreseeable future.

Gibb (Chapter 11) seeks to give substance to hopes for greater regional integration by highlighting ways in which southern Africa could benefit from European experience. He concentrates upon theories of integration, in particular those based on interpretations of the 'federalist' and 'functionalist' intellectual traditions concerned with European integration in the immediate aftermath of World War II. Both seek ultimately to establish a new world order by reducing sovereign powers of nation states, but functionalists adopt a pragmatic approach establishing a network of overlapping institutions, each with a specific task or function. Action in one or more areas then builds up pressure for integration in other spheres, leading to a gradual erosion of national sovereignty. The European Coal and Steel Community and the Single European Act are advanced as examples of this neo-functionalist, 'spillover' approach.

Within southern Africa, Gibb focuses on the nature and evolution of SADC as the organization most likely to promote regional economic integration. SADCC initially rejected models of economic integration based on trade liberalization, in favour of an approach based on project coordination, but this failed to promote closer regional political co-operation or economic integration. In the early 1990s, SADCC moved towards a 'developmental integrative approach' which Gibb shows to bear close resemblance to that enshrined in the 1957 Rome Treaty. This implies some acceptance of free trade and economic liberalization, co-ordination of macroeconomic policies and limited pooling of sovereign powers. In part such changes must be seen as a response to implications of impending majority rule in South Africa.

The 1992 SADC treaty established a framework of co-operation that prioritizes deeper economic co-operation and integration, common economic and political systems and a common foreign policy, all before an effectively functioning trade area has been established. Gibb points out that SADC is promoting policies normally associated with a later stage of trade integration—in effect the European federalist strategies pursued after World War II—and argues that SADC should learn from Europe the advantages of adopting the neo-functionalist spillover strategy, co-operating in specific areas such as tourism, energy and the environment.

The immensity of the tasks facing South Africa's new government will be obvious from all the foregoing. At home it must seek at least partially to accommodate the great expectations of its own people, conscious that failure to do so may threaten both political stability and ultimately its own hold on government. Such domestic pressure may make it difficult to prioritize regional initiatives, yet the ANC will be conscious of its debt to neighbouring countries for their support during the liberation struggle, and will in any event wish to normalize relations between South Africa and the rest of the region. Third, the eyes of the world will be on South Africa. Its success in gaining aid and investment will depend upon assessment of its internal stability and economic performance by Western governments, bankers and businessmen. At a more profound level, the world will be looking to South Africa's new rulers to create a peaceful, nonracial democracy on the unpromising foundations of apartheid.

REFERENCES

Crankshaw, O. (1993) 'Squatting, apartheid and urbanisation on the southern Witwatersrand', *African Affairs*, 92(366), 31–51.
Crankshaw, O. and Hart, T. (1990) 'The roots of homelessness: causes of squatting in the

Vlakfontein settlement south of Johannesburg', *South African Geographical Journal*, 72(2), 65–70.

Christopher, A.J. (1994) *The Atlas of Apartheid*, Routledge, London.

de Klerk, W.A. (1975) *The Puritans in Africa: A Story of Afrikanerdom*, Rex Collings, London.

Furnivall, J.S. (1948) *Colonial Policy and Practice*, Cambridge University Press, Cambridge.

Heymans, C. and Totemeyer, G. (eds) (1988) *Government by the People?*, Juta, Cape Town.

Huntington, S.P. (1981) 'Reform and stability in a modernizing, multi-ethnic society', *Politikon: South African Journal of Political Science*, 8(2), 8–26.

Lemon, A. (1976) *Apartheid: a Geography of Separation*, Saxon House, Farnborough.

Lemon, A. (1982) 'Migrant labour and frontier commuters: reorganizing South Africa's Black labour supply' in D.M. Smith (ed.), *Living Under Apartheid*, George Allen and Unwin, London, 64–89.

Lemon, A. (1987a) *Apartheid in Transition*, Gower, Aldershot.

Lemon, A. (1987b) 'Swaziland' in C.G. Clarke and A. Payne (eds), *Politics, Security and Development in Small States*, George Allen and Unwin, London, 156–69.

Lemon, A. (ed.) (1991) *Homes Apart: South Africa's Segregated Cities*, Paul Chapman, London.

Lemon, A. (1992) 'Restructuring the local state in South Africa: Regional Services Councils, redistribution and legitimacy' in D. Drakakis-Smith (ed.), *Urban and Regional Change in Southern Africa*, Routledge, London, 1–32.

Mabin, A. (1989) 'Struggle for the city: urbanisation and political strategies of the South African state', *Social Dynamics*, 15(1), 1–28.

Pickles, J. and Woods, J. (1992) 'South Africa's homelands in the age of reform: the case of QwaQwa', *Annals of the Association of American Geographers*, 82(4), 629–52.

Republic of South Africa (1993) *Constitution of the Republic of South Africa Bill*, B212-93(GA), Government Printer, Pretoria.

Smith, D.M. (1990) *Apartheid in South Africa* (Update series), 3rd edition, Cambridge University Press, Cambridge.

Smith, D.M. (ed.) (1991) *Urbanization in Contemporary South Africa: the Apartheid City and Beyond*, Unwin Hyman, London.

Urban Foundation (1990) *Policies for a New Urban Future: Urban Debate 2010: 2 Policy Overview: the Urban Challenge*, Urban Foundation, Johannesburg.

Western, J. (1981) *Outcast Cape Town*, George Allen and Unwin, London.

Wills, T.M. (1991) 'Pietermaritzburg' in A. Lemon (ed.), *Homes Apart*, 90–103.

Postscript: The Election and Beyond

ANTHONY LEMON

In April 1994, South Africa held its first truly 'general' election. The election was only the third in this century to produce a change of government, a result which, unlike those of 1924 and 1948, was never in doubt. It was nevertheless greeted with joy by most South Africans, and with worldwide acclaim. Within the country, the sense of liberation following decades of oppression and struggle was overwhelming. Globally, the election set the seal upon a process which has little historical precedent: a racial oligarchy had negotiated the transfer of power to the population as a whole.

The months approaching the election were dominated by attempts to induce both the black and white right, bizarrely united in the Freedom Alliance, to participate, in the hopes of ending violence. Success was achieved in several stages (Lemon, 1994): first the Ciskeian government announced its participation; in Bophuthatswana a recalcitrant President Mangope was overthrown in early March 1994, and his would-be rescuers from the Afrikaner Weerstandsbeweging humiliated, an event which catalysed the break-up of the white right and the formation of General Constand Viljoen's Freedom Front to fight the election and support the cause of a *Volkstaat*, or Afrikaner homeland. This left Chief Buthelezi and his Inkatha Freedom Party as the main threat to a fair and peaceful election. Inkatha's participation was eventually secured just one week before the election, after a series of public concessions, including constitutional entrenchment of the status of the Zulu monarchy, and a deal kept secret from the ANC until after the election whereby control over all land in KwaZulu was transferred from the bantustan government to the King, with the apparent intention of denying control over the Zulu heartland to any post-election ANC regional government.

During the election campaign, many areas of KwaZulu-Natal were effectively no-go areas for parties other than Inkatha. The ANC too was guilty at local level of trying to exclude other parties from campaigning in several areas (Reynolds, 1994). With threats from the black and white right effectively removed, however, the election itself was remarkably free of violence, despite a multitude of procedural shortcomings and some malpractice. There was almost 'a hush of expectation, and a quiet in the air' as the emotional nature of the occasion and all it symbolized overcame the frustrations and discomforts of the shambolic voting conditions which prevailed in many areas. The turnout was high, with some 88 per cent of an estimated 22.7 million electors voting.

In five of the nine regions the ANC won more than three-quarters of all votes cast. It won more narrowly in the Pretoria–Witwatersrand–Vereeniging (PWV) region, and with just under half the total poll in the Northern Cape. The National Party won the Western Cape, where it reduced the ANC share of the poll to just under one-third. Official results gave the ANC a similar share in KwaZulu-Natal, with Inkatha gaining a narrow overall majority (50.3 per cent). This result was the outcome of negotiations following substantial evidence of electoral malpractice in the region; it was disputed by the regional ANC leadership, but agreed in the wider national interest by a national leadership aware that an ANC administration might find KwaZulu-Natal ungovernable and anxious to avoid protracted disputes, violence and negative overseas reactions which could have disastrous effects on investment.

Even if official polling figures are accepted, the geographical limits to Inkatha's support are clear. Its greatest strength was in the rural areas, especially in northern KwaZulu, whereas the ANC gained substantial majorities in Durban and Pietermaritzburg. Inkatha received few votes in those eastern areas in dispute between KwaZulu-Natal and the Eastern Cape (see Chapter 2). The same was true in those areas of the Eastern Transvaal where Zulu-speakers are numerically dominant.

Remarkably, the NP appears to have gained half its total support, or even slightly more, from groups which it treated as second-class citizens or non-citizens under apartheid. It probably won the support of the Indian and coloured communities by as large a margin (60–70 per cent) as it won that of whites, with the result that about 37 per cent of NP support was drawn from these groups (Reynolds, 1994). A further 14 per cent came from black voters, representing 3–4 per cent of the black vote, a small but significant bridgehead into the black community on which the NP will hope to build in the next election (Reynolds, 1994 p. 193).

The NP victory in the Western Cape depended on support from coloured voters, whose fears of affirmative action and African dominance in the ANC it successfully exploited in the election campaign. In rural areas, however, more coloured people appear to have supported the ANC, perhaps because coloureds and Africans are less clearly separated spatially and occupationally in these areas; thus the NP was deprived of the victory it hoped for in the rural Northern Cape. Indians had less impact than coloured voters on the outcome of the election, owing to their smaller numbers (3 per cent of the electorate) and geographical concentration in Natal, where their allegiance appears to have divided approximately two to one in favour of the NP over the ANC (Reynolds, 1994).

Nationally, the ANC's 62.6 per cent share of the poll fell just short of the critical two-thirds majority required to pass the final constitution to be implemented in 1999, a result which was greeted with relief by President Nelson Mandela who stressed that a final constitution, like the interim constitution, should be the product of negotiation. The National Party's 20.4 per cent share of the poll was just above the critical threshold to secure a Deputy Executive Presidency, while Inkatha's national share of 10.4 per cent was sufficient to bring the party three cabinet posts in the new government of national unity, including the appointment of Buthelezi as Minister of Home Affairs.

Despite the use of proportional representation, minority parties with a proud record—the liberal Democratic Party and radical Pan Africanist Congress—gained minimal representation, leaving their long-term future in doubt. Both were essentially squeezed in an election where voting was largely on racial lines, except in KwaZulu-Natal where Zulu ethnicity was successfully mobilized by Inkatha. The PAC suffered from a virtually invisible leader, from bankruptcy, and more profoundly from having let its Africanist philosophy degenerate into anti-white racism. The DP's problems were well summed up by Allister Sparks (1994, p. 10): 'the DP failed to adjust to the changed political environment. It had established itself and built up its reputation as a party of protest against an oppressive system, but at the critical moment of transition it shrank from identifying with the black struggle to replace that system'. Its approach appeared strident at times, and its liberalism came increasingly to be identified with free market ideology.

The Freedom Front won 14 per cent of the estimated white vote nationally, but 22 per cent in the provincial elections. By splitting the white right, General Viljoen effectively put paid to any serious hopes of para-military opposition to the new government. Neither the level of Freedom Front support nor its distribution (its best performances were in the PWV, Orange Free State and Northern Cape regions, where it won about 6 per cent of the poll) were helpful to the cause of a *volkstaat*, but they were sufficient for the white right to retain a parliamentary voice in these regions and nationally.

A peaceful election involving all South Africans is a huge achievement, but South Africa is only half-way through its transition (Esterhuysen, 1994). The election has produced governments of national unity at central and regional levels which are to govern for five years, during which time the country's final constitution is to be written. Consolidation of the transition to democracy demands commitment from all those elected to national reconciliation and integration, and success in tasks of post-apartheid reconstruction and economic development.

In mid-1994, as South Africa began to confront these tasks, conditions were relatively favourable in some respects: the economy was emerging from recession, inflation remained under 10 per cent, interest rates were scheduled to decline as capital investment flows rendered balance of payments constraints less onerous, and renewed global economic expansion promised to raise prices for the country's major mineral exports. The ending of apartheid had created a situation in which international capital markets were prepared to lend, and foreign companies returned in significant numbers in 1994. With the ending of sanctions, old export markets are once more open and new ones beckon, for instance in the export of armaments. Membership of the Southern African Development Community became a reality in September 1994, heralding a new era in South Africa's political and economic relations with its neighbours.

There are, however, many problems. Borrowing has dangerous implications for interest rates, unless the funds acquired are used to expand the productive base. Budget savings and redirected spending to fund the government's modified Reconstruction and Development Programme are planned to rise from R2.5 billion in 1994–5 to R10 billion in ten years time, but it is doubtful that such savings can be achieved: even defence spending, cited by Nelson Mandela and others as an area of potential saving during the election campaign, looks likely to rise in the first instance, given the costs of integrating liberation armies and those of 'independent' homelands into the new South African National Defence Force. Security expenditure is also likely to rise, to combat the problems of crime, violence and urban dislocation.

Substantial aid was promised soon after the election, notably by Japan and the USA which announced packages of R4.7 billion and R2 billion respectively, but this must be seen in the perspective of a social backlog estimated by the World Bank at R46 billion. In the short term at least, foreign investment must be set against substantial capital outflows—R3.2 billion in the months January–April 1994—in part reflecting white attempts to secure their future outside the country should they eventually decide to leave. Nor does all foreign investment produce jobs: much of that which occurred in 1993 and early 1994 involved little more than a distribution agreement or a buy-back into an existing operation (Davie, 1994).

Even supposing a broadly favourable economic climate for reconstruction, there are formidable obstacles of a non-economic nature which are likely to impede progress. At national level, numerous issues could strain relations between those now sharing power (Esterhuysen, 1994). These include central–provincial relations, the search for truth about past political atrocities and amnesty for the perpetrators, the implementation of affirmative action, and the need for financial stability and fiscal discipline. The greatest danger arises from the inevitable resentment of sections of the population which receive no early benefits from political change; it will be remarkable if the ANC, hitherto a popular revolutionary movement, can persuade people of the need to wait three to five years or more for tangible evidence that their needs being met. The difficulty of maintaining the ANC's pre-election alliance with the trade unions was quickly underlined in the months after the election with the occurrence of strikes and government resistance to wage demands which are not backed by improved productivity.

Regional and local government also face daunting problems. The new provincial administrations must amalgamate areas previously administered by several separate authorities (see Chapter 2), including former provinces, homelands and national 'own affairs' departments of the House of Representatives and

House of Delegates. This represents a formidable organizational task, the implementation of which will inevitably delay delivery of new policies. Even the location of regional capitals remained uncertain in some cases, notably KwaZulu-Natal where Inkatha's wish to govern from the KwaZulu capital of Ulundi was fiercely contested by other parties which wished to retain the Natal provincial capital of Pietermaritzburg (*Weekly Mail*, 1994). In the Eastern Transvaal the debate was more regionally based, with Witbank and Nelspruit the leading contenders from the highveld and lowveld regions of the province respectively. In both cases the compromise of a split capital might be adopted, but this would be an example of what van Zyl Slabbert has described as 'governments of national unity ... spending to pay for their compromises' (*Star International Weekly*, 1994).

Reform of local government is at an even earlier stage in many areas. The Local Government Negotiating Forum was launched only in 1993, and local negotiations between local authorities made variable progress in different parts of the country during 1993 and 1994. The Reconstruction and Development Programme envisaged that 800 old racially constituted local authorities would eventually be replaced by 300 non-racial ones. Here too a situation of constitutional transition at the local metropolitan level will affect the capacity of the local state to deliver the improvements which are widely expected of it.

Southall (1994) has characterized the election as having given South Africa a 'one-party dominant' government, suggesting parallels with other African states. The resounding victory of the ANC in the 1994 election will make it hard to displace in the foreseeable future. The existence of a more developed civil society than elsewhere in Africa, together with the countervailing power of commerce and industry, give cause to hope that its dominant position will not lead the ANC to become arrogant and authoritarian. Such an assured position should help the ANC to govern with its eyes on a longer-term perspective than most governments elected for five-year terms, resisting populist policies in the interests of a more fundamental and enduring reconstruction of a 'new South Africa'.

REFERENCES

Davie, K. (1994) 'Footing the bill', *Towards Democracy*, 3(2), 12–16.
Esterhuysen, P. (1994) 'Comment: towards one-party dominance in South Africa?', *Africa Insight*, 24(2), 82–5.
Lemon, A. (1994) 'A balance of forces: the new South Africa', *The Month*, 27(6), 220–25.
Reynolds, A. (ed.) (1994) *Election '94 South Africa*, James Currey, London.
Southall, R. (1994) 'South Africa's 1994 election in an African perspective', *Africa Insight*, 24(2), 86–98.
Sparks, A. (1994) *Weekly Mail*, Johannesburg, 16–22 June, p. 10.
Star International Weekly, Johannesburg (1994), 28 April–4 May, p. 14.
Weekly Mail, Johannesburg (1994) 16–22 June, p. 1 and 14–20 July, p. 4.

Part I

GETTING THERE: CONSTITUTIONAL ISSUES

1 Post-apartheid South Africa and the Nation-State

A.J. CHRISTOPHER

The concept of the nation-state continues to occupy a central place in political geographical studies. Attention has been directed towards the frequent lack of spatial correspondence between state and nation (Glassner, 1993; Taylor, 1993). Consequently a growing number of governments have had to address the problem of reconciling conflicting ethnic aspirations within their national borders. The outcomes have been variable, but multi-national states have proved to be remarkably fragile in the 20th century, as the rapidly expanding literature on state partition and secessionism has indicated (Buchanan, 1991; Gams, 1991; Horowitz, 1992; Premdas *et al.*, 1990; Waterman, 1984, 1987). The destruction of the three Slavic federations (the Soviet Union, Yugoslavia and Czechoslovakia) in the early 1990s has given added impetus to such studies, as state disintegration rather than integration has become the norm. In the period from the 1920s to the 1970s the heritage of British colonial 'divide and rule' policies was particularly significant in many of the problematical state partitions of the present century, including that attempted in South Africa (Christopher, 1988). The end of empire, of whatever nature, thus appears to be a particularly disruptive era, as previously repressed peoples attain or strive to attain political freedom.

Geographical research has focused on the problems leading to and following partition, as well as seeking general guidelines for assessing the 'success' of breakaway states. Factors including external pressures, internal political accommodation, ethnic concentration and population transfers have influenced the long-term viability of the partition. External pressures have determined the outcome of several partitions, whether as the result of defeat in war (Pakistan), outside military intervention (Cyprus) or the major powers abstaining from intervention (Palestine). The lack of internal political accommodation has usually been the factor determining the conversion of an autonomist into a secessionist movement (Sri Lanka). By modifying internal structures, political compromise has had the opposite effect of persuading dissident groups to retain allegiance to the existing state (Spain). The spatial distribution of the ethnic minorities desiring secession is significant. Ethnic compaction within an historic territory has proved to be important for 'success', as spatially compact groups, often situated on the periphery of the state, have been most likely to succeed (Slovenia). Scattered minorities have been less successful, unless substantial population transfers take place to fit the population to the political map (Bosnia-Herzegovina). Although in

The Geography of Change in South Africa, edited by Anthony Lemon.
© 1995 by the Editor and Contributors. Published in 1995 by John Wiley & Sons Ltd.

the final analysis force, or the threat of force, has usually been essential to attain secession, some have been achieved by mutual agreement (Czechoslovakia). On the other hand repression may be of such an intensity as effectively to destroy the secessionist movement (Tibet).

APARTHEID IN SOUTH AFRICA

Between 1959 and 1991 the South African government pursued a state partition policy aimed at establishing ten additional separate independent sovereign states. In devising this policy the government drew upon existing colonial and other models but made little attempt to disguise the real intent of the policy, namely the maintenance of exclusive white power over most of the state territory. Constitutionally only four new 'independent' states were in being when the re-establishment of the single state became official policy at the end of 1991. The basis of the partition was the creation of separate states (homelands) for a number of linguistically defined African ethnic groups, with the white population retaining control of the remainder of South Africa (Christopher, 1994). Parallels with the multi-ethnic Soviet and Yugoslav nationalities policies are evident, rather than the bipolar partitions most common since 1945. However, the similarity was not pursued further as the equitable distribution of resources, notably land, was not contemplated.

The policy was expounded to sceptical Western governments in terms of the decolonization of Africa (*Times*, 6 April 1959; *New York Times*, 26 April 1959). The political transformation of the continent in the late 1950s and early 1960s radically changed the moral position of the white minority government in South Africa. Harold Macmillan, the British Prime Minister, delivered his 'Winds of Change' speech to the South African parliament in Cape Town in 1960, immediately before the country left the Commonwealth and so symbolically heralded a new era (*Times*, 4 February 1960).

The South African government, however, was determined that even limited political freedom for the African population was not to infringe upon white dominance. The problem thus bore little relationship to the classic role of minority-group aspirations recently outlined by Mikesell and Murphy (1991) or the development of minority ethnonationalism described by Hennayake (1992). Pressures for ethnic autonomy or secession in such cases has come from the minorities, denied effective access to political power. Majority or governing groups have usually opposed such moves until a reversal of policy became politically expedient. In South Africa the powerless African majority did not seek either autonomy or partition but participation in national politics. The partition scheme was devised by the governing ethnic minority group with the purpose of retaining power. Partition was offered as a solution to the conflict between two apparently politically incompatible philosophies of government within a single state, namely white minority rule and African majority rule. The minority government was determined that such a policy was to inconvenience the white population as little as possible and hence the results appeared to be a continuation of white rule.

The territorial basis of the new states rested on racial patterns of land allocation (see below). The ethnic basis of the states was elaborated later, in an attempt to foster minority nationalisms. The failure to do so was paralleled by the failure to establish legitimacy for the new states and their governments. Gross economic inequalities between the governing group and the fostered nationalities compounded the failure. However, the potential for partition has not disappeared with the end of apartheid, although the concept of ethnic partition has been discredited in the eyes of most blacks as a result of its association with apartheid.

THE STATE

The ethnic basis of the partition of South Africa was not to be translated into a straight division of the land between the indigenous African population who predominated in the eastern two-thirds of the country and the white and mixed-race coloured who predominated in the western section. Such a division would inevitably have resulted in the acquisition of the industrial heartland of the Witwatersrand by the Africans, as theoretical partition plans have suggested (Blenck and von der Ropp, 1977; Maasdorp, 1980). Earlier suggested academic plans had retained a section of the Transvaal for the whites, but these proposals were quickly shown to be untenable (Jooste, 1991). The lack of any white majority territorial base resulted in the South African partition policy differing markedly from other 20th-century state partitions. Even the western, non-African, third of the country registered a 'coloured' majority, which was politically unacceptable to the National Party, then intent on excluding this community from access to political representation.

Instead, the pattern of landholdings determined the territorial basis of partition. The cadastre not the census was to determine state areas, a situation only paralleled by the partition of Palestine in the 1940s (Stein, 1984). Thus the area set aside as the Native Reserves and Trust Lands under legislation enacted in 1913 and 1936 became the basis of the 10 new nation-states or African homelands. Under these two measures some 13 per cent of the area of South Africa was designated for exclusive African occupation and the remainder was to stay under white control. The homeland territories thus consisted of a series of fragmented blocks. Accordingly, much of the government's attention in the 1960s and 1970s was directed towards 'consolidation' to reduce the number of territorial outliers belonging to the various states (see Figure 1.1). Until the 1970s the government opposed any increase in the overall area to be assigned to the homelands and

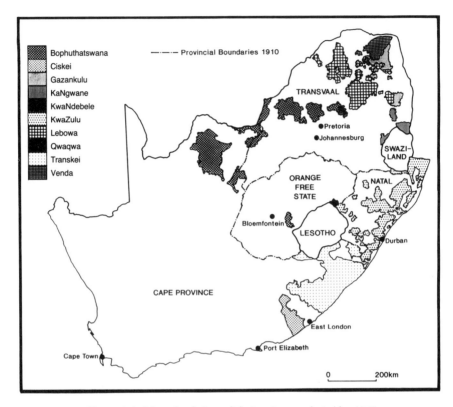

Figure 1.1. Homeland Consolidation Proposals in the 1970s

consolidation merely involved the accelerated purchase of land designated in 1936 for the extension of the African areas, a process only completed in the 1980s. Consolidation was a high priority of homeland governments, recognizing the unfavourable comment of foreign governments and industrialists upon the multiplicity of territorial outliers and inliers. However, without a radical redrawing of the homeland boundaries to include extensive tracts of white-owned land, no such consolidation was possible.

The new African homeland governments quickly reproduced the traits of independent states elsewhere in Africa. Conflict in the form of territorial disputes arose, without the restraining influence of the Organization of African Unity. Significantly the majority of disputes were based on historic claims, not ethnicity. In this manner they paralleled the wider international experience since the Second World War (Murphy, 1990). Thus the Transkei government sought to re-establish its boundaries to coincide with those of the Transkeian Territories in the early colonial period through the incorporation of land which had subsequently passed into white ownership. Such claims were rare and not successful.

The South African government initially assumed that Botswana, Lesotho and Swaziland could be co-opted into the new state system. Maps prepared by the Tomlinson Commission in 1955 indicated that these states would be extended to incorporate lands presently within South Africa (South Africa, 1955). The Botswana government rejected the proposals. That of Lesotho was more intent on regaining lands in the eastern Orange Free State (the 'Conquered Territories') lost in the 1860s than acquiring impoverished Qwaqwa. Only the Swaziland government appeared interested in the acquisition of South African territory on the terms offered. A preliminary agreement was reached in 1982 providing for the transfer of the KaNgwane homeland and the Inguavuma district of KwaZulu to the kingdom (Griffiths and Funnell, 1991). Under the agreement large tracts lost by the Swazi state in the 19th century would have been regained and Kosi Bay acquired as a potential port. However, the death of King Sobhuza II of Swaziland in the following year, together with the opposition of the KwaZulu and KaNgwane governments, effectively quashed the plan, and the internationally recognized boundaries of South Africa remained unaltered.

THE NATION

In order to avoid confronting a united and numerically superior African nation, each of the African languages was assumed to identify a separate group or nation (Figure 1.2). Many of the divisions were the result of comparatively recent standardizations of the spoken language. Thus modern Xhosa was standardized by the missionaries at Lovedale in the western area and subsequently enforced within the eastern Cape Colony. Modern Zulu was standardized using the speech prevalent in the area north of the Tugela. Consequently two linguistic groups sharing a common language origin had their languages standardized on the two dialects most distant from one another. In practice the resulting division may not be as clearcut as official parlance would suggest. This was well illustrated by the recent (1992) marriage of the Zulu king to a Xhosa princess. In response to the media reaction that this represented a highly desirable Zulu–Xhosa *rapprochement*, it was pointed out by the KwaZulu government that the bride came from a Zulu tribe which had resided on the other (Cape) side of the colonial border and hence had always, incorrectly, been described as Xhosa (*Times*, 27 July 1992). Similar, often arbitrary, linguistic reclassifications took place elsewhere (Vail, 1989).

The problems of defining the numerical viability of the nation were not so immediately solved. Until the 1970s the north and south Ndebele, with under 500 000 members, were regarded as being too small and scattered to support a separate national structure. However, linguistic pressures exerted upon the

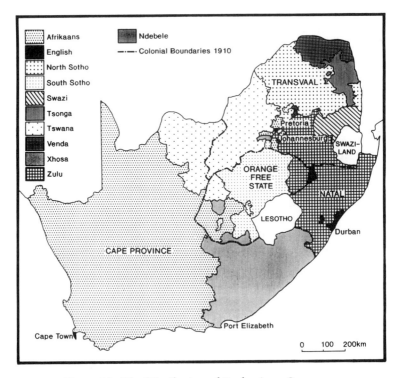

Figure 1.2. The Distribution of Predominant Languages

Ndebele by the homeland governments, more particularly Bophuthatswana and Lebowa, engendered a sense of ethnic persecution and demand for a separate political status, if not for independence (James, 1990). In contrast the perpetuation of historic divisions within the colonial administration of the Xhosa peoples led to the establishment of two Xhosa-based states (Anon., 1989).

At the commencement of the implementation of the homeland policy, the majority of the African population of South Africa resided outside its designated nation-states (Christopher, 1982). In an effort to make the homelands closer to the concept of the nation-state, the government resorted to large-scale population removals and resettlements. Between 1960 and 1983 at least three and a half million people were moved to fit the apartheid plan (Platzky and Walker, 1985; Unterhalter, 1987). The majority were related to the homeland policy. Some one and a half million Africans were moved from white-owned farms, in an attempt to create a 'whiter' rump state. A further 1 100 000 were removed from African-owned land and from small reserves considered to be poorly situated for consolidation into the homelands. Many of the 600 000 urban removals were concerned with the transfer of urban Africans to the homelands. Thus the policy involved population transfers amounting to approximately one-sixth of the 1970 population of the country—'ethnic cleansing' by another name.

Parallel with the movement of Africans from the rump white state to the homelands was the transfer of people between homelands to fit the nation-state ideal. Thus South Sotho people were moved from the Bophuthatswana enclave of Thaba Nchu to adjacent Botshabelo, which was originally intended to be incorporated into Qwaqwa (Murray, 1992). Even Xhosa-speaking Ciskeians were moved from Transkei to Ciskei after the political transfer of the Glen Grey and Herschel districts between the two states (Cobbett and Nakedi, 1988). Conflict over the Moutse district between KwaNdebele and Lebowa in the mid-1980s formed part of the wider struggle against the former's 'independence' movement. Significantly, the coercive power necessary to effect and maintain this transfer began to break down in the later 1980s with the result that Africans moved back over the homeland boundaries into 'white South Africa' in considerable numbers (Kane-Berman, 1990).

The rump South African state consequently retained an African, and therefore theoretically alien, majority throughout the period that partition was state policy. In 1992 approximately a third (10 million) of the African population of the country still resided within 'white' South Africa in comparison to five million whites. Even including the four million people classified as 'Asian' and 'coloured', who had been unwillingly co-opted into the South African constitutional framework in 1984, potential 'foreigners' formed the majority of the population. Extensive areas adjacent to the homelands contained a majority of the ethnic group of that homeland, a serious geopolitical point of weakness. Movement across the boundaries was easy and control was often minimal. The potential for instability was therefore profound.

The linguistically complex area of the Witwatersrand is the hub of the South African economy. It lies outside the core territory of any one African linguistic group. Thus those attracted to the job opportunities in the region came from all over South Africa and indeed beyond. The African townships accordingly contained a highly complex population, which the government sought to attach politically and emotionally to the homelands. Hence ethnic zoning was introduced in the townships to enable mother-tongue instruction to be offered in schools, and cultural and other links maintained within the groups rather than merge into a larger South African nation. In practice ethnic residential integration did take place and political co-operation ensued (Christopher, 1989).

RESOURCES

The resources and basic infrastructure of the pre-partition South African state remained in white hands through the manipulation of the new boundary lines.

Thus the main lines of communications passed along white corridors. On the Zimbabwean border the section of Venda adjoining the Limpopo River was expropriated, enabling the South African army to retain security control. African suburbs in white towns were incorporated, where possible, into the homelands leading to a massive daily 'international' commuter flow of nearly one million to work (Lemon, 1982). Most significantly few white towns, and none of any size, were incorporated into the homelands. In the quest to hold as much territory as possible, the white government retained extensive areas where few whites lived. The security zones on the Zimbabwean and Botswanan borders were virtually devoid of white farmers, who were only induced to live there with the promise of substantial state subsidies.

The most damning evidence against the partition programme was the inequality of the distribution of resources between the homelands and the rump 'White South Africa'. In 1985, although 41.6 per cent of the total population of South Africa now lived in the homelands, they accounted for only 4.7 per cent of gross domestic product (GDP). The resultant figures for GDP per head in the homelands reveal the semi-destitute character of the population. Thus figures of R250 to R780 per head in the homelands compare with a figure of R6260 in 'White South Africa' (Development Bank of Southern Africa, 1993). Bophuthatswana recorded the highest figure due to the revenues from the platinum mines in the country. The large-scale resettlement of essentially destitute people without the attendant provision of employment was responsible for the low figure for KwaNdebele and several other homelands. The homelands were thus not taken as the basis of administrative units by any of the significant parties to the constitutional negotiations (see Chapter 2).

LEGITIMACY

Establishing a degree of legitimacy was vital for the governments of the homelands. State constitutions provided for the inclusion of a large body of appointees within the legislatures. The Transkei constitution, the role model for homeland political development, provided for half the National Assembly to be nominated by the governing party. The government was assured of a parliamentary majority, even though it failed to win electoral approval in the pre-independence polls. In Ciskei, an enquiry into the popular response to independence was undertaken in 1980 (Ciseki, 1980). Despite the negative reaction documented in the ensuing report, this did not deter the Ciskeian government from taking independence the following year.

The promotion of KwaNdebele as the fifth independent state in the mid-1980s failed as the result of internal dissension in a society composed almost entirely of 500 000 refugees from other parts of South Africa, particularly from Lebowa and Bophuthatswana (Ritchten, 1989). During the late 1980s the strains within the system intensified (Peires, 1992). In 1987 the military staged the first of two *coups d'état* to overthrow the Transkei government. The following year the Bophu-thatswana government was only saved from a similar fate by the intervention of the South African army. In 1990 the Venda and Ciskei governments were overthrown by their respective armies. All three military governments formed links with South African political movements, notably the African National Congress, in opposition to the South African government. Subsequently the Ciskei military government drew closer to the South African government, coming into conflict with the Transkei government and hence the African National Congress. When the Conference for a Democratic South Africa (CODESA) met for the first time in December 1991 only the Bophuthatswana government wished to continue the independent homeland system, although the KwaZulu and subsequently Ciskei governments sought a high degree of autonomy for the enlarged areas they assumed they would dominate.

The combined factors of hostile external pressures and internal unacceptability invalidated the policy of partition (Lemon, 1991). No country outside South Africa recognized any of the homelands as sovereign entities. Furthermore, a policy based on race and involving state fragmentation incurred intense inter-national hostility, notably from the Organization of African Unity and the United Nations. South Africa by the 1980s had thus become one of the 'pariah states', with severe international consequences of restricted diplomatic contacts and the imposition of economic sanctions (Geldenhuys, 1990). The universal rejection of a racially based policy was such that the terms for the re-admission of the country into the international community were the establishment of a non-racial, but also a *united*, South Africa.

The attempted partition of South Africa was a highly contentious issue involving the legalization of racial discrimination with the international con-demnation which went with it. The policy of partition was narrowly conceived and often brutally enforced. The majority therefore remained firmly opposed to it as few advantages were evident for the African population. The policy of state partition failed as the African majority became more influential in the affairs of state, particularly as a result of the endemic violence and economic stagnation which overcame the country in the late 1980s.

With the formal end of apartheid, new partition proposals by white right-wing organizations provided for a continuing white state within a predominantly African-controlled southern Africa. It is significant that the areas for these proposed white states bore little relationship to one another and all lacked demographic viability (du Toit, 1991; *Sunday Times*, [Johannesburg] 11 July 1993). The northern Transvaal, the sole region in which the white electorate rejected the re-establishment of the unitary state in the March 1992 referendum, could not be regarded as the core of a white nation-state as only 5 per cent of the population would have been citizens! In contrast the demand for the secession of the ethnic core of the coloured population in the western third of the country raises more fundamental problems parallel with the era of apartheid (*New Nation*, 14–22 May 1993).

FUTURE PARTITION PROSPECTS

It may therefore be asked 'does the concept of state partition have any place in the politics of post-apartheid South Africa?' The experience of the international community in recent years would suggest that the answer may be 'yes'. Clearly the application of the concept is now unrelated to the nationally and internationally discredited attempt by the white minority to secure a separate state. Prospects for a future partition must therefore be viewed against the background of more orthodox historically and territorially based movements. What follows is offered not as a political programme but as an academic speculative warning as to the potential for fragmentation.

The Organization of African Unity since its foundation has been successful in diverting attention away from the fragility of the African state system and so averting its disintegration. So successful has the Organization's programme been that it has been suggested that some deviation may be possible from the rigid adherence to the doctrine of not allowing any boundary changes on the continent (Cobbah, 1988; Deng, 1991; Mayall and Simpson, 1992). In 1993 Eritrea was the first special case to gain an internationally recognized independence, portrayed as the ending of Ethiopian colonial rule. The re-establishment of a dubiously erased colonial boundary presents a parallel with the Western Sahara, and thus poses no serious threat to the stability of the continental state system. The current *de facto* separation of Somaliland (ex-British) from Somalia (ex-Italian) may prove to be permanent. The case of the southern Sudan is more problematical as the region never achieved a formal separate colonial identity, although the administrative regimes operating in the two sections of the former Anglo-Egyptian condominium were significantly different. However, the division of some African states may be the only way to establish political stability and an end to the wastage of valuable and limited state resources on internal repression. In this respect African leaders may have to re-examine the territorial implications of 'self-determination' from which they have shielded themselves since independence (Young, 1991).

The four provinces of South Africa were separate colonial entities prior to the establishment of the Union in 1910. Both the Transvaal and Cape of Good Hope are ethnically heterogeneous and may be dismissed as potential coherent political entities. The African populations of the Orange Free State and Natal are more homogeneous and so warrant more serious attention. Within the Orange Free State, South Sotho separatism or Basotho irredentism are not in evidence. The prospect of the re-emergence of an independent Orange Free State on to the international scene or the creation of a 'greater' Lesotho are not contemplated, although the incorporation of Lesotho into a post-apartheid South Africa has been suggested (Cobbe, 1991). Furthermore the central geographical situation of the Orange Free State militates against secession, which has traditionally had greater chances of success in peripheral situations.

Natal, the most politically distinctive and coherent province, may prove to be the unit for which a case for secession could be presented (Forsyth and Mare, 1992). The resultant bipolar partition of South Africa would be based not on the white–black or immigrant–indigene models, which have dominated political thinking in the post-1948 era, but on the self-determination of the Zulu nation, the largest ethnic group in the country, within a clearly defined colonial boundary. However, the coincidence between state and nation would not be exact as the resultant Zulu state of Natal would include substantial white and Indian minorities (approximately 22.5 per cent of the total) and exclude extensive areas in the south-eastern Transvaal and eastern Orange Free State housing approximately 19.0 per cent of the Zulu nation. The KwaZulu government has demanded self-determination and secession if its political requisites are not met in the drafting of the new South African constitution (*Times*, 5 October 1992). Furthermore it appears unlikely that the area involved would be limited to the fragmented homeland but would include the whole of the historic colony and province of Natal (*Times*, 5 December 1992).

Natal has been politically distinctive from the remainder of South Africa during the last two hundred years. Prior to the imposition of colonial rule, the strife which accompanied the rise and fall of the Zulu military monarch resulted in the emergence of a distinct national group with a recognizable language and culture (Morris, 1985). The British colonial society which emerged in the second half of the 19th century was equally violent, as exemplified by its response to the Bambata rebellion, causing Winston Churchill to describe Natal as 'the hooligan of the British Empire' (Hyam, 1968, p. 251). After union in 1910, the English-speaking white electorate of Natal was usually in opposition to the Afrikaner majority of the electorate in the remainder of the country.

Crisis point was reached after the republican referendum in 1960, when the secession of the province was temporarily placed on the political agenda (Thompson, 1990). Subsequently, the KwaZulu government and the Inkatha Freedom Party, which controls it, have deliberately pursued a policy aimed at fostering Zulu nationalism and giving that nation a sense of its own history and separate identity (Forsyth, 1992; Mare, 1992). This has led to somewhat eccentric views of South African history. The statement by one senior Inkatha Freedom Party official that 'we are the Serbs of South Africa', provides warning enough (*Weekly Mail*, 21–7 May 1993).

Nationalism within a post-apartheid South Africa will have to be fostered carefully, while Zulu ethnic identity may not be compatible with the wider loyalty. The conversion of an autonomist nationalism into a secessionist movement demanding independence has been a significant feature in the development of a number of states (Orridge and Williams, 1982). The exact trigger mechanism for this conversion is remarkably ill-defined. In South Africa, African solidarity, however fragile, has been attained in opposition to apartheid and the domination of the white government. However, in the process the majority of the population has been made aware of its ethnic origins through separate ethnic education systems and the promotion of indigenous languages. Once political struggle is no longer viewed as liberation from white domination, then the disruptive spectre of ethnically based politics appears to be probable (Horowitz, 1991).

In the late 1980s and early 1990s Zulu society was riven by a sanguinary conflict between the supporters of the African National Congress and the minority Inkatha Freedom Party (Adam and Moodley, 1992; Bekker, 1992; Taylor and Shaw, 1994). Significantly, politics were still viewed by both sides within a South African context. However, if at some future date, Zulu political leaders were to seek 'Zulu self-determination' and secession instead of accommodation, as alluded to by King Goodwill Zwelithini, a parallel with the situation in 1948 might arise (*Times*, 13 July 1993). The pertinent difference is that the Zulu nation occupies an historic territorial base, a condition which the whites so significantly lacked in 1948. On the other hand, the KwaZulu government and the Inkatha Freedom Party lack the political power to impose their solution upon the remainder of the population, a power which the whites did possess in 1948. The Zulu leadership may, however, have the power to disrupt the country as many other ethnic-based movements on the African continent have done since 1960. This raises the speculation that South Africa might be able to achieve a more prosperous and peaceful future without the destabilizing need to coerce a potentially hostile and disruptive population in Natal. Parallels with the separation of Pakistan from India, Singapore from Malaysia or southern Ireland from the United Kingdom may be drawn. In these cases, the acquiescence of the majority in the secession of the minority ensured a reduction in internal conflict and the international stability of the outcome. Paradoxically a special constitutional provision recognizing the Zulu right to secede, and hence attain ethnic security, has been suggested as the means of preventing partition (Buchanan, 1991). Otherwise the country may face the pessimistic prophecy by Adam and Moodley (1993, p. 219) that 'Natal . . . and the western Cape . . . could emerge as the Croatia and Slovenia of South Africa.'

CONCLUSION

The exploitation of ethnicity for political purposes has played a highly destructive role in the modern evolution of South Africa. The failure of the white population to attain a numerical majority over the indigenes, as was attained in other mid-latitude British dominions, or to establish any historic majority region, doomed the attempt to retain political control over even a portion of the country. No partition plan which ignored both African aspirations and those of the Asian and coloured communities could overcome the minority status of the white population throughout the country. Furthermore the partition plans drawn up by the National Party government were implemented with considerable brutality and injustice towards the African majority. Parallels with decolonization were therefore bogus. Ethnicity is thus a suspect political issue for the majority of African politicians in the post-apartheid era. However, governmental and academic failure to recognize its continuing significance has been highly disruptive for many states in recent years. In the transitional period to a post-apartheid era, South African politics have become a matter of achieving ethnic co-operation rather than separation. However, the potential for severe political conflict, even secessionism, remains extremely high in what is by any standards an exceptionally volatile multi-ethnic state.

REFERENCES

Adam, H. and Moodley, K. (1992) 'Political violence, "tribalism", and Inkatha', *Journal of Modern African Studies*, 30 (3), 495–510.

Adam, H. and Moodley, K. (1993) *The Opening of the Apartheid Mind: Options for the New South Africa*, University of California Press, Berkeley.

Anon. (1989) 'Ethnicity and pseudo-ethnicity in the Ciskei' in L. Vail (ed.), *The Creation of Tribalism in Southern Africa*, James Currey, London, 395–413.

Bekker, S. (1992) *Capturing the Event: Conflict Trends in the Natal Region 1986–1992*, Indicator South Africa, Durban.

Blenck, J. and von der Ropp, K. (1977) 'Republic of South Africa: is partition a solution?', *South African Journal of African Affairs*, 7 (1), 21–32.

Buchanan, A. (1991) *Secession: the Morality of Political Divorce from Fort Sumter to Lithuania and Quebec*, Westview Press, Boulder.

Christopher. A.J. (1982) 'Partition and population in South Africa', *Geographical Review*, 72 (2), 127–38.

Christopher, A.J. (1988) '"Divide and rule"—the impress of British separation policies', *Area*, 20 (3), 233–40.

Christopher, A.J. (1989) 'Apartheid within apartheid: an assessment of official intra-Black segregation on the Witwatersrand, South Africa', *Professional Geographer*, 41 (3), 328–36.

Christopher, A.J. (1994) *Atlas of Apartheid*, Routledge, London.

Ciskei (1980) *Report of the Ciskei Commission*, Conference Associates, Pretoria.

Cobbah, J.A.M. (1988) 'Toward a geography of peace in Africa: redefining sub-state self-determination rights' in R.J. Johnston, D.B. Knight and E. Kofman (eds), *Nationalism, Self-determination and Political Geography*, Croom Helm, London, 70–86.

Cobbe, J. (1991) 'Lesotho: what will happen after apartheid goes?', *Africa Today*, 38 (1), 18–32.

Cobbett, W. and Nakedi, B. (1988) 'The flight of the Herschelites: ethnic nationalism and land dispossession' in W. Cobbett and R. Cohen (eds), *Popular Struggles in South Africa*, James Currey, London, 77–89.

Deng, F.M. (1991) 'Dilemmas of nation building: racism, ethnicity and development in Africa', *Ethnic Studies Report*, **9**(2), 1–9.

Development Bank of Southern Africa (1993) Unpublished Gross Domestic Product statistics.

du Toit, B.M. (1991) 'The far right in current South African politics', *Journal of Modern African Studies*, **29** (4), 627–67.

Forsyth, P. (1992) 'The past in the service of the present: the political use of history by Chief A.N.M.G. Buthelezi 1951–1991', *South African Historical Journal*, **26**, 74–92.

Forsyth, P. and Mare, G. (1992) 'Natal in the New South Africa', *South African Review*, 6 (358), 141–51.

Gams, I. (1991) 'The Republic of Slovenia: geographical constants of the Central European state', *GeoJournal*, **24** (4), 331–40.

Geldenhuys, D. (1990) *Isolated States: a Comparative Analysis*, Jonathan Ball, Johannesburg.

Glassner, M.I. (1993) *Political Geography*, John Wiley, New York.

Griffiths, I. and Funnell, D.C. (1991) 'The abortive Swazi land deal', *African Affairs*, **90** (358), 51–64.

Hennayake, S.K. (1992) 'Interactive ethnonationalism, an alternative explanation of minority ethnonationalism', *Political Geography*, **11** (6), 526–49.

Horowitz, D.L. (1991) *A Democratic South Africa? Constitutional Engineering in a Divided Society*, University of California Press, Berkeley.

Horowitz, D.L. (1992) 'Irredentas and secessions: adjacent phenomena, neglected connections', *International Journal of Comparative Sociology*, **33** (1–2), 118–30.

Hyam, R. (1968) *Elgin and Churchill at the Colonial Office, 1905–08*, Macmillan, London.

James, E. (1990) 'A question of ethnicity: Ndzundza Ndebele in a Lebowa village', *Journal of Southern African Studies*, **16** (1), 33–54.

Jooste, C.J. (1991) 'Partition as a constitutional option' in D.J. van Vuuren, N.E. Wiehahn, N.J. Roodie and M. Wiechers (eds), *South Africa in the Nineties*, Human Sciences Research Council, Pretoria, 227–57.

Kane-Berman, J. (1990) *South Africa's Silent Revolution*, South African Institute of Race Relations, Johannesburg.

Lemon, A. (1982) 'Migrant labour and frontier commuters: reorganizing South Africa's Black labour supply' in D.M. Smith (ed.), *Living under Apartheid: Aspects of Urbanization and Social Change in South Africa*, George Allen & Unwin, London, 64–89.

Lemon, A. (1991) 'Apartheid as foreign policy: dimensions of international conflict in southern Africa' in N. Kliot and S. Waterman (eds), *The Political Geography of Conflict and Peace*, Belhaven Press, London, 167–83.

Maasdorp, G. (1980) 'Forms of partition' in R.I. Rotberg and J. Barratt (eds), *Conflict and Compromise in South Africa*, David Philip, Cape Town, 107–50.

Mare, G. (1992) *Brothers Born of Warrior Blood: Politics and Ethnicity in South Africa*, Ravan, Johannesburg.

Mayall, J. and Simpson, M. (1992) 'Ethnicity is not enough: reflections on protracted secessionism in the Third World', *International Journal of Comparative Sociology*, **33** (1–2), 5–25.

Mikesell, M.W. and Murphy, A.B. (1991) 'A framework for comparative study of minority-group aspirations', *Annals of the Association of American Geographers*, **81** (4), 581–604.

Morris, D.R. (1985) *The Washing of the Spears: the Rise and Fall of the Great Zulu Nation*, Abacus, London.

Murphy, A.B. (1990) 'Historical justification for territorial claims', *Annals of the Association of American Geographers*, **80** (4), 531–48.

Murray, C. (1992) *Black Mountain: Land, Class and Power in the Eastern Orange Free State 1880s to 1980s*, Edinburgh University Press, Edinburgh.

New Nation (Johannesburg).

New York Times (New York).

Orridge, A.W. and Williams, C.H. (1982) 'Autonomist nationalism: a theoretical framework for spatial variations in its genesis and development', *Political Geography Quarterly*, 1, 19–39.

Peires, J.B. (1992) 'The implosion of Transkei and Ciskei', *African Affairs*, 91 (364), 365–87.

Platzky, L. and Walker, C. (1985) *The Surplus People: Forced Removals in South Africa*, Ravan, Johannesburg.

Premdas, R.R., Samarasinghe, S.W.R. de A., Anderson, A.B. (1990) *Secessionist Movements in Comparative Perspective*, Pinter, London.

Ritchten, E. (1989) 'The KwaNdebele struggle against independence', *South African Review*, 5, 426–45.

South Africa (1955) *Summary of the Report of the Commission of Enquiry for the Socio-economic Development of the Bantu Areas within the Union of South Africa*, Report U.G.61-1955, Government Printer, Pretoria.

South Africa (1992) *Population Census 1991, Home Language*, Report No. 03-01-06, Government Printer, Pretoria.

Stein, K.W. (1984) *The Land Question in Palestine, 1917–1939*, University of North Carolina Press, Chapel Hill.

Sunday Times (Johannesburg).

Taylor, P.J. (1993) *Political Geography: World-economy, Nation-state and Locality*, Longman, London.

Taylor, R. and Shaw, M. (1994) 'The Natal conflict' in J.D. Brewer (ed.), *Restructuring South Africa*, Macmillan, Basingstoke.

Thompson, P.S. (1990) *Natalians First: Separatism in South Africa 1909–1961*, Southern Books, Johannesburg.

Times (London).

Unterhalter, E. (1987) *Forced Removal: the Division, Segregation and Control of the People of South Africa*, International Defence and Aid Fund for Southern Africa, London.

Vail, L. (ed.) (1989) *The Creation of Tribalism in Southern Africa*, James Currey, London.

Waterman, S. (1984) 'Partition—a problem in political geography' in P.J. Taylor and J.W. House (eds), *Political Geography: Recent Advances and Future Directions*, Croom Helm, London, 98–116.

Waterman, S. (1987) 'Partitioned states', *Political Geography Quarterly*, 6 (2), 151–70.

Weekly Mail (Johannesburg).

Young, C. (1991) 'Self-determination, territorial integrity, and the African state system' in F.M. Deng and I.W. Zartman (eds), *Conflict Resolution in Africa*, Brookings Institution, Washington, 320–46.

2 Regional Proposals: their Constitutional and Geographical Significance

RODDY FOX

The debate over regions during the 1991–3 negotiating process is of critical consequence to our understanding of future developments, both political and economic, in South Africa. Major political subdivision of national territories frequently produces striking disparities in development indicators, and South Africa, through apartheid, has exemplified this phenomenon. The negotiations that have ensued at CODESA and the Multi-Party Negotiating Process provide an opportunity to examine the regional dispensations proposed by the African National Congress (ANC), the National Party (NP) and the three reports by the Commission for the Demarcation/Delimitation of Regions (CDDR). Particular attention will be paid in the sections below to the criteria used by these organizations to delimit regions. Different stances with regard to electoral systems, federal and national powers will be examined as will their relationships to the geographically unequal distribution of ethnic groups and wealth.

REGIONAL DISPENSATIONS

INITIAL PROPOSALS

The NP's initial stance was that the nine Development Regions, used for industrialization and development policies through the 1980s, be used as a starting point for negotiation (Humphries, 1992). Figure 2.1 shows these nine regions in relation to the Provinces, TBVC States (Transkei, Bophuthatswana, Venda, Ciskei) and Self-Governing Territories which were the creations of grand apartheid. One reason for the use of the nine regions was that they had become increasingly entrenched in the country's administrative structure since they were introduced at the Good Hope Conference in the early 1980s. The nine regions have been viewed by both of the major players in the negotiation process, the NP and ANC, as having the advantage of technical neutrality (National Party, 1993; African National Congress, 1993); they were initially drawn up using planning criteria. Figure 2.2 portrays the ANC's original 10-region proposal from their 1992 discussion document (Constitutional Committee of the African National

The Geography of Change in South Africa, edited by Anthony Lemon.
© 1995 by the Editor and Contributors. Published in 1995 by John Wiley & Sons Ltd.

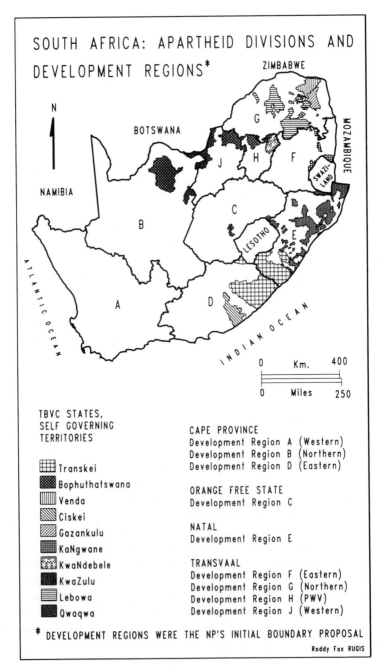

Figure 2.1. Homelands and Development Regions

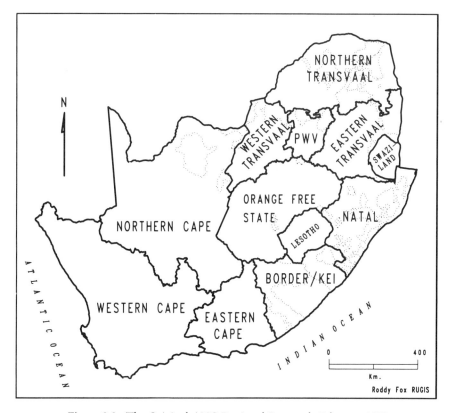

Figure 2.2. The Original ANC Regional Proposal, February 1992

Congress, 1992), and the influence of the Development Regions on ANC thinking is clear.

Regional development documents from the 1980s (Republic of South Africa, 1988, pp. 5–6) describe the planning criteria used as regional delimiters, particularly the nodality principle which assessed:

- Most important nodes and their sphere of influence;
- Geographical distribution of the population;
- Natural resources;
- Existing infrastructure;
- Physical characteristics and topographic features in particular;
- Economic activities; and
- Statistical and political boundaries. . . .

. . . the following criteria were also considered:

- Demarcation within a southern African context;
- Homogeneity in respect of development needs, development potential and physical characteristics;
- Functional relationships;
- Ethnicity;
- Identifiable regional cores; and
- Associated metropolitan areas.

Later sections will show that many of these criteria were also used by the ANC (African National Congress, 1993) and CDDR (Commission on the Demarcation/Delimitation of SPRs, 1993a) in their deliberations. The regions drawn up through implementing these criteria (Figure 2.1) used the existing provincial boundaries to delimit the Orange Free State (Region C) and most of Natal (Region E), a feature in common with virtually all of the other dispensations that will be discussed below. The Cape and Transvaal Provinces were split into Western, Northern and Eastern portions and the Transvaal was further divided by the Pretoria–Witwatersrand–Vaal triangle (PWV). Generally speaking this left one TBVC State/Self-Governing Territory in each Development Region. However, some of the bantustans were split between Development Regions: Transkei fell into Regions D and E, Bophuthatswana in Regions B, C, H and J (Figure 2.1). The result of overlaying these functional planning regions on the mosaic of states created through grand apartheid has been the triplication and quadruplication of second-level government throughout the country. Tables 2.1 and 2.2 demonstrate the multiplications which have arisen through this system. It will be seen later that the regional delimitations presented by the ANC and CDDR have attempted to overcome this splitting of ethnic bantustans, particularly in the new Eastern Cape and the North-west Provinces. Thus the spatial imprint of apartheid influenced the delimitation process.

Table 2.1. Administrative Subdivision of Bantustans

Bantustan	Separate Divisions	Development Regions
Transkei	3	D,E
Bophuthatswana	7	B,C,H,J
Venda	3	G
Ciskei	1	D
Gazankulu	3	G
Kangwane	3	F
KwaNdebele	1	H
KwaZulu	21	E,F
Lebowa	12	G
Qwaqwa	1	C

Source: compiled by author.

Table 2.2. Administrative Subdivision of Development Regions

Development Region	Separate Divisions	Separate 'Governments'
Region A (W. Cape)	1	1
Region B (N. Cape)	3	2
Region C (OFS)	3	3
Region D (E. Cape)	4	3
Region E (Natal)	25	3
Region F (E. Transvaal)	5	3
Region G (N. Transvaal)	20	4
Region H (PWV)	4	3
Region J (W. Transvaal)	3	2

Source: compiled by author.

NP thinking about boundaries and regions shifted notably with the introduction of a seven-region proposal in September 1992 (Figure 2.3) which was first displayed at their August regionalism conference (Development Action Group, 1992). The nine regions were reduced to seven in response to three developments: the ANC's proposals for 10 regions (Constitutional Committee of the African National Congress, 1992), the Northern Cape/Western Transvaal conference of April 1992 (*Beeld*, 1992), and the NP's potential electoral prospects in alliance with conservative black homeland leaders. This latter prospect seemed likely before such leaders as President Mangope, Brigadier Gqozo and Chief Minister Buthelezi formed the Freedom Alliance and its precursor COSAG (the Concerned South Africans Group), later in 1993.

Figure 2.3 shows the Cape of Good Hope region as an expanded Western Cape whose borders have moved to the north, to the Orange River, and east, to include the Port Elizabeth metropole. The North-west region consisted of the Northern Cape beyond the Orange River, Western Transvaal and four of Bophuthatswana's seven divisions. This version of the North-west was an early NP attempt to combine Bophuthatswana into one region; later it will be shown that more territory was included in further dispensations to include five of the divisions. The Transvaal region included Pretoria, which was excised from the PWV, another theme which will be returned to later, and the Northern and Eastern Transvaal Development Regions. Kei region was Transkei, Border, Ciskei and that part of Eastern Cape Development Region D to the east of Port Elizabeth. Natal and the Orange Free State were left virtually untouched with the exception of the inclusion of Umzimkulu into Natal. Umzimkulu is the north-easterly of the two detached portions of Transkei, which appears as an island surrounded by KwaZulu and Natal in Figure 2.1. The problem of the positioning of Umzimkulu was another recurrent theme in the delimitation process. The ANC, for example, placed it within the Border/Kei region in their February 1992 proposal (Figure 2.2). An important influence on these seven regions may well have been the

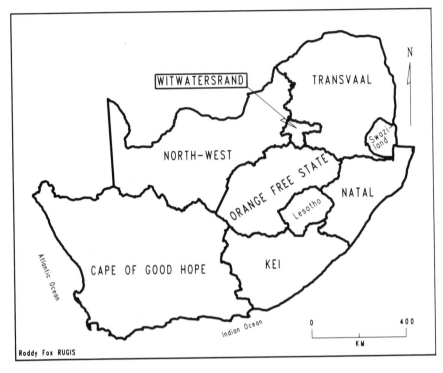

Figure 2.3. The NP Seven-region Proposal, September 1992

language distribution of the country. *The Language Atlas of South Africa* (Grobler *et al.*, 1990) shows clearly dominant languages in four of these regions since the regional boundaries proposed were very close to the language distributions: the Cape of Good Hope (Afrikaans), North-west (Tswana), Kei (Xhosa) and Natal (Zulu).

The ANC's discussion document, *Ten Proposed Regions for a United S.A.* (Constitutional Committee of the African National Congress, 1992) was another influence on the National Party's seven-region proposal. Before examining its impact on NP thinking, it is worth examining the ANC's delimitation criteria in some detail as they bear comparison with the Development Region criteria listed earlier and they were consistently applied in the ANC's two main proposals.

There were six main criteria and three further considerations. First, regions should be economically and socially functional taking into account the tension between functionally interdependent regions which may be large and the need for setting achievable planning and administrative goals which may well require smaller regions. Second, regions should allow for 'balanced' rural and urban development. Presumably this means that development of both rural and urban sectors should be possible, thus redressing the plight of the roughly 50 per cent of the black population living in rural areas. Third, regions should be as geographically compact as possible, avoiding 'the fragmentation, terrestrial peninsulas and isolated enclaves which racial gerrymandering has brought to our country' (Constitutional Committee of the African National Congress, 1992, p. 13). This criterion clearly influenced the ANC's thinking with regard to Umzimkulu, incorporating it with Mount Currie (a part of Natal Province) into the Border/Kei region. In later dispensations it also clearly influenced the boundaries of the North-west Region. Fourth, equality of income between regions should be a realistic goal. Thus the size, population and geographic product of the regions should be borne in mind in the delimitation process. Fifth, administrative boundaries, whether Provincial, District or Magisterial, should be used to establish the exact regional borders but only where people were accustomed to them and where they did not have racist or ideological connotations. Sixth, current regions with popular acceptance and a 'sense of rightness' should be considered. The ANC went on to say that the nine Development Regions had this type of acceptance and were therefore used as a basic framework. There were, however, three important qualifications to these criteria.

The first was that Development Regions D and E split Transkei into two and: 'This might or might not have made sense from a purely economic point of view but from any other perspective would be manifestly unviable. No one regards northern Transkei as a natural or organic part of Natal' (Constitutional Committee of the African National Congress, 1992, p. 13). Accordingly, they reverted to the pre-1910 frontier of the Cape Province for the Border/Kei region. The 1910 boundaries were their second qualification, to be used to rectify what they saw as the anomalous position of Mafikeng/Mmabatho, which they returned to the Northern Cape from Western Transvaal, and Sasolburg, which they moved from the PWV to the Orange Free State. The 1910 boundaries were presumably thought to be free from racist connotations. The third qualification was that there was a need for a mechanism to be put in place to enable consultation with people in border areas over minor boundary adjustments.

Given that the ANC's criteria were broadly similar to those used to create the Development Regions, it was not surprising that their proposals were similar. The main difference was over the reamalgamation of Transkei by the ANC.

The NP's subsequent seven-region proposal appears to have used the principle implicit in the argument quoted above concerning Transkei's artificial subdivision. The proposal has a Kei region and a North-west region which would reamalgamate much of Bophuthatswana with the Northern Cape and Western Transvaal. It could easily have been argued that an alliance with President Mangope would enhance the NP's electoral prospects among black voters, who were in a large majority (Table 2.3) in this region.

Some of the major players in the Northern Cape/Western Transvaal/Bophu-

Table 2.3. Wealth Distribution and Population in CDDR Regions, 1991

Province	Income per Capita 1985 (R)	Asian (%)	Black (%)	Coloured (%)	White (%)	Population ('000)
PWV	4575	1.6	70.6	3.2	24.6	9267.2
KwaZulu/Natal	1971	11.1	80.0	1.5	7.5	7590.2
Eastern Cape	1360	0.2	87.2	6.7	5.9	6137.0
Northern Transvaal	725	0.1	97.1	0.1	2.7	4525.3
Western Cape	4373	0.7	17.8	59.7	22.1	3392.7
Orange Free State	2184	0.0	85.2	2.7	12.1	2723.3
North-West	2000	0.3	86.5	2.8	10.4	2396.6
Eastern Transvaal	2341	0.5	84.6	0.9	14.0	2129.5
Northern Cape	2817	0.2	31.0	53.9	14.9	726.6
SOUTH AFRICA	2580	2.7	75.6	8.7	13.0	38 888.4

Source: Commission on the Demarcation/Delimitation of SPRs (1993a).

thatswana region had also mooted the idea of a North-west region (Drummond, 1991; *Beeld*, 1992) with the convening of regional forums in March 1991 and April 1992. The forums consisted of prominent academics in Politics and Planning Departments at Potchefstroom University and the University of Bophuthatswana, the Bophuthatswana government, regional interest groups, political parties, and white right-wing farmers. They agreed upon the boundary of a new region (Figure 2.4) whose population was deemed to be Tswana and Afrikaans speakers. These forums seem to have lobbied successfully as the region was now shown on the NP's seven-region proposal (Figure 2.3) and subsequent CDDR delimitations.

The Democratic Party's *Discussion Paper on Constitutional Proposals* (Democratic Party, 1992) deals in part with regions, and it is worth examining briefly because of the honest-broker role the DP played at CODESA and the Multi-Party Negotiating Process. The DP's regions were very much a compromise between the NP and ANC positions. The DP favoured a federation of ten states: the nine Development Regions and a state consisting of the Ciskei, Border, Transkei area. They also proposed considering the merging of parts of the Northern Cape and Western Transvaal.

The bantustan leaders also expressed a number of ideas about their borders (Humphries and Shubane, 1992). For example, President Holomisa of Transkei and Brigadier Gqozo of Ciskei laid claims to various parts of the Eastern Cape,

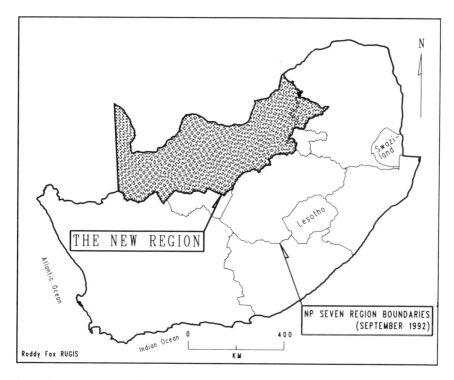

Figure 2.4. Proposed Northern Cape, Bophuthatswana, Western Transvaal Region, April 1992

Border, and also various portions of each other's territory (*Eastern Province Herald*, 9 July 1993).

REGIONS, ELECTORAL SYSTEMS AND THE CONSTITUTION

Delimiting regions becomes very important when strong powers are devolved to that level of government. The geographical distribution of voters, when each race group exhibits different party allegiances, and the distribution of wealth also become two critical issues.

DISTRIBUTION OF THE ELECTORATE AND OF WEALTH

The geographical distribution of white voters has been examined by Fox (1992). The white vote potentially consisted of just under four million voters, 16.1 per cent of the electorate. Of these voters 60 per cent lay in the country's four major metropoles: the PWV (39 per cent); Cape Town (11 per cent); Durban–Pinetown (six per cent); Port Elizabeth (four per cent). The National Party's white support, therefore, lay principally in the PWV and Cape Town. The much smaller rural vote from the Afrikaner platteland was likely to go to the Conservative Party, a Freedom Alliance member. An additional attraction to the NP was the fact that the PWV and Cape Town were the country's two wealthiest regions (Fox, 1994) as the income per capita figures in Table 2.3 indicate. The Western and Northern Cape regions also housed the bulk of the country's 'coloured' population and polls showed significant support for the National Party in these regions (*Sunday Times*, 1994). Thus these two regions, whose white and coloured populations combined are 81.8 and 68.8 per cent of the total (see Table 2.3), and the PWV, with a white population 25 per cent of the total, were of great electoral significance to the NP.

The black population consisted of between 70 and 97 per cent of the population in the remaining regions of the country. With one notable regional exception, consistently high support levels for the ANC/SACP alliance of between 58 and 65 per cent were polled among the black population (de Kock *et al.*, 1993; *Sunday Times*, 1994). The exception was the Zulu supporters of the Inkatha Freedom Party (IFP) who constituted an estimated 10 per cent of the total black vote. These supporters lived largely in Natal and the PWV; thus an alliance between the NP and Inkatha could have been expected to have regional support in both the PWV and Natal. Such electoral prospects would have been reinforced if the Indian community, largely found in Natal, were also to have supported the NP or Inkatha.

NP regional, constitutional and electoral thinking was fundamentally affected by their need to broaden support in the black groups. Given that these groups were not equally distributed throughout the country (Table 2.3), then the NP at times looked towards regional alliances. Thus they were trying in 1992 to establish a National Front—an alliance with coloured MPs and conservative black homeland figures such as President Mangope, Chief Minister Buthelezi and Brigadier Gqozo. It is with these alliances in mind that their seven-region proposal should be assessed. It should have garnered support in the proposed Cape of Good Hope (with its predominance of 'coloured' Afrikaans speakers), the North-west (the Tswana, Afrikaans region) and Natal (the Zulu region), while attempting to split the ANC's vote through the support of Brigadier Gqozo in the Kei Region (the Xhosa region). As mentioned earlier, it is interesting that these four regions were really the only ones which bear comparison with ethnic regions in South Africa.

The NP's subsequent submission to the CDDR, detailed below, shows that they had moved away from this line of thinking by 1993. Presumably, this shift came after the founding of COSAG, the Concerned South Africans Group which later became the Freedom Alliance, by black and white conservative interests.

These alliances lay to the ideological right of the National Party and held out for a confederal system.

The distribution of wealth was another key factor in the debate over the boundaries and powers of regions. Under a federal system with wealth 'locked' within Provincial borders it would be hard for an ANC government to transfer development funding either from region to region or within regions which it did not control. South Africa's Development Regions were marked by disparities between the wealthiest regions, the PWV and Western Cape; a middle tier of regions, the Eastern Transvaal, Orange Free State, Western Transvaal, Natal; and the poorest regions, the Eastern Cape, Northern Cape, and Northern Transvaal. Furthermore, there were additional disparities between the white and black spatial components within these regions (Fox, 1994).

The need to redress the unequal distribution of wealth was one of the ANC's goals in the demarcation process and a key influence on the lobbying to the CDDR for the apportionment of Pretoria in either Northern or Eastern Transvaal since both regions were much poorer than the PWV (Commission for the Demarcation/Delimitation of SPRs, 1993b). It was also a feature in the debate over having separate Eastern Cape and Border/Kei regions. Business and other interests in the Port Elizabeth area strongly voiced the opinion that they should not be responsible for the improvement of Border/Kei—their poor neighbour—such uplifting should come from national funds.

ELECTORAL SYSTEMS AND THE CONSTITUTION

The two major players in the negotiation process, the ANC and NP, started the negotiation process favouring a unitary state and a federal state, respectively. Humphries and Shubane (1992) summarized the tensions and shifts within ANC policy but concluded, in line with their discussion document (Constitutional Committee of the African National Congress, 1992), that the ANC was committed to elected regional authorities. Regional powers could either be specifically stated in the constitution or given in more general terms and be subject to periodic review by the national legislature.

The ANC's initial legislative proposal was for a 400-member assembly with 200 members selected from regional lists and 200 from national lists. Proportional representation was to be used at both regional and national levels, and so the delimitation of regions was not all-important. The ANC's proposal was subsequently ratified by the Multi-Party Negotiating Process (Wessels, 1993) and provided the basic framework for the 27 April 1994 elections. In fact there was a good deal of consensus from the President's Council Report (Republic of South Africa, 1992) and the DP (Democratic Party, 1992) on the use of the proportional representation system and the number of seats.

The ANC initially favoured using the original four Provinces, presumably with reincorporated TBVC States and Self-Governing Territories, as the regional basis for election to the National Assembly (*Weekly Mail*, 1992; *Daily Dispatch*, 1992). Subsequent ANC thinking (African National Congress, 1993) showed that they were in favour of using the country's new regions, delimited by the CDDR, for electoral purposes but felt that the National Assembly, to be elected on 27 April 1994, was the body to arrive at final boundary resolutions.

In contrast to the ANC, the NP was trying to establish the right of minority parties to block legislation at the regional and national level (Humphries and Shubane, 1992). Thus the NP proposed an upper house of regional representatives, with the right of veto over legislation from the lower house. The regional delegations were to consist of an equal number of members from the parties which achieved a certain minimum number of votes. The NP failed in this attempt since the composition of the Senate adopted by the Multi-Party Negotiating Process consisted of 10 members from each of the nine CDDR regions, the 10 members to be elected by the regional legislature in proportion to the voting in that region.

The Senate does not have any unusual blocking powers (Wessels, 1993; Multi-Party Negotiating Process, 1993) over ordinary legislation which can be carried by a simple majority. The Senate may not amend any bills relating to taxation or revenue; however, bills affecting provincial boundaries and the exercise of provincial powers and functions must be approved by a joint sitting of the National Assembly and the Senate. Perhaps of most importance is the clause dealing with amendments to powers allocated to provinces. Such amendments have to be passed by a two-thirds majority in the National Assembly and the Senate, 'provided that the legislative and executive competencies of a province shall not be amended without the consent of its legislature' (Wessels, 1993, p. 9). The interim constitution also specified that a Commission on Provincial Government be appointed to finalize the number, boundaries, powers and functions of the Provinces. The Commission would also streamline the setting up of regional governments.

The interim constitution allowed for extensive powers for the regions, although these could be modified through the mechanisms described above: the joint sitting of National Assembly and Senate or the Commission on Provincial Government. The powers were concurrent with Parliament's right to make laws and covered: agriculture; casinos, racing, gambling and wagering; cultural affairs; primary and secondary education but not university/technikon education; health services; housing; language policy and the selection of regional official languages; local government, subject to national provisos; nature conservation but not nationally administered bodies like the parks, botanical gardens, marine resources; police, subject to national provisos; provincial public media; public transport; regional planning and development; road traffic regulation; roads; tourism, trade and industrial promotion; traditional authorities; urban and rural development; welfare services.

THE SECOND ROUND OF REGIONAL PROPOSALS

THE COMMISSION ON THE DEMARCATION/DELIMITATION OF STATES/PROVINCES/REGIONS

The original proposals of the ANC, NP, other political parties and interest groups (McCarthy, 1993; Pienaar, 1992) were publicized and debated through 1992. In mid-1993 the Multi-Party Negotiating Process asked for formal presentations to the CDDR (Commission on the Demarcation/Delimitation of SPRs, 1993a).

The CDDR was established at the end of May 1993 by the Negotiating Council, and it was empowered to report back with proposed boundaries in only six weeks. The Commission received 304 written submissions and listened to 80 oral presentations in Johannesburg, Cape Town, Port Elizabeth, Durban and Umzimkulu. The Commission consisted of 15 members, two of whom were to decline to sign the CDDR's report and presented minority reports, and there was a technical support team of 12 members.

There were ten delimiting criteria which the CDDR were to use:

- historical boundaries (Provincial, Magisterial and District) and infrastructure;
- administrative considerations such as the presence or absence of infrastructure and nodal points for services;
- the need to rationalize existing apartheid creations (the TBVC states etc.);
- the need to limit financial and other costs as much as possible;
- the need to minimize inconvenience to people;
- the need to minimize the dislocation of services;
- demographic considerations;
- economic viability;
- development potential;
- cultural and language realities.

The Commission aggregated these into four groups: economic aspects, geographical coherence, institutional and administrative capacity, socio-cultural issues. It is pertinent that this list of criteria, and their subsequent grouping, were similar to the ANC's criteria and also to those used to delimit the original nine Development Regions.

From the outset there were a number of problems with these criteria and the context within which they were placed. First, and perhaps most important, was the fact that the boundaries of the regions (CDDR jargon was States/Provinces/ Regions to placate political sensitivity to the issue) were to be demarcated before the powers of the regions were known. This was a critical issue since it would impact on the size of the regions. In the Eastern Cape, for example, including Port Elizabeth and its hinterland with Border/Kei would undoubtedly have increased the region's revenue base. But this was only really important if the bulk of development funding had to come from within the region. Should development funding come from national coffers then it would probably be more sensible in terms of cultural, language and administrative criteria to have a separate Eastern Cape and Border/Kei.

A similar problem bedevilled the incorporation of Kruger National Park into one or another of the two regions in the Transvaal. Resolving the problem depended on whether the Park was to be managed as a national asset or whether its control devolved to the regional level—this was not known at the time of demarcation.

Another big drawback of having these 10 criteria was that their application was bound to cause contradictions. Thus an economically functional boundary could, and in a number of cases did, cut across a cultural or administrative boundary. Additionally, there was no deadlock-breaking mechanism built into the CDDR process to allow the Commission to decide which criteria were to be applied where there was conflict. Consequently, a number of contentious areas arose in the country. Associated with this problem was the vagueness of some of the criteria. How did they measure demographic considerations, economic viability or the need to minimize inconvenience? The regions produced would depend to a large measure on how these were assessed, and there would be a number of ways in which this could be done.

THE ANC'S SUBMISSION

The criteria listed by the ANC were much the same as those described earlier for their 10-region proposal. Perhaps of more significance was their mission statement: '. . . all bantustans must be reincorporated into South Africa and a united country created. There is a need for active movement away from apartheid, overcoming: the enormous fragmentation and inequalities, the harmful ethnic divisions, and the regional imbalances' (African National Congress, 1993, p. 2). By implication, therefore, there would be attempts to include different race groups within one region where this did not impact on other criteria. The ANC's submission also showed that they were probably looking to local revenue as a source of development funding. Thus in comparison with Figure 2.2, Figure 2.5 shows a coalesced Eastern Cape and Border/Kei (Region D), a coalesced Northern Cape and Western Transvaal (Region B) and an expanded Western Cape (Region A). In each of these cases the ANC was annexing a poorer area with a richer area. In the Eastern Cape, the ANC was also attempting to amalgamate the predominantly black population of two different bantustans, Transkei and Ciskei, with the Border corridor and Port Elizabeth area in which were found the majority of the region's white and coloured population. The ANC's other regions remained the same as their 1992 proposal.

The ANC document also discussed internal shifts in ANC policy as the party

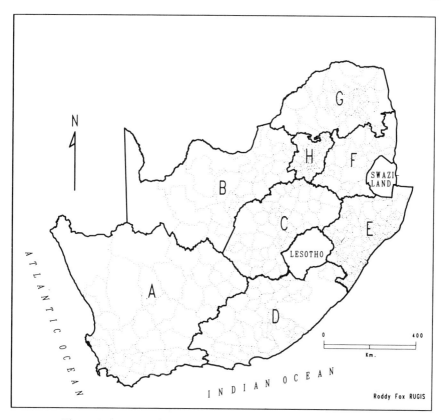

Figure 2.5. Regional Proposal Submitted by the ANC, July 1993

examined first the 10-region map (Figure 2.2) and then a 16-region map before deciding that they would seek ten regions or fewer. Significantly, the ANC viewed the regions as being electoral creations and considered that their borders should be flexible, pending decisions by the National Assembly.

THE NATIONAL PARTY'S SUBMISSION

Figures 2.6 and 2.7 show the two scenarios which the National Party proposed (National Party, 1993). They preferred scenario A, a nine-region proposal, but also submitted scenario B with seven regions should nine regions be too many. The seven-region proposal simply amalgamated the four regions in the Cape Province into two.

Both maps showed that National Party thinking had now altered in some fairly radical ways. Natal was the only region proposed in September 1992 which was

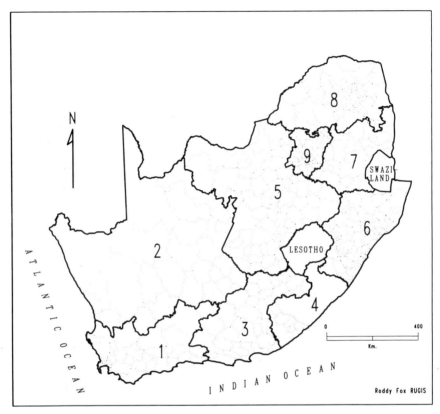

Figure 2.6. Nine-region Proposal Submitted by the NP, July 1993

still to be found in the same configuration in the July 1993 proposals. Perhaps the most remarkable change was in the amalgamation of the Orange Free State with the North-west. In commenting on this amalgamation and other changes the NP made the questionable assertion that, 'should the way of life and organisation patterns of the civic community be taken into account, it is evident that the population of the various proposed regions portray a regional identity throughout' (National Party, 1993, p. 5). Elsewhere (p. 16) they somewhat more pragmatically say that as much of Bophuthatswana as possible has been put into this region. Thus the North-west had to be expanded to include the Orange Free State within which lies the Bophuthatswanan enclave of Thaba Nchu (Figure 2.1). At the same time they excised from the North-west the Bophuthatswanan districts of Odi 1 and Moretele 1 which lay within the borders of the PWV. This was justified on functional grounds since there were strong economic linkages between these two districts and the PWV. It has also been noted elsewhere that

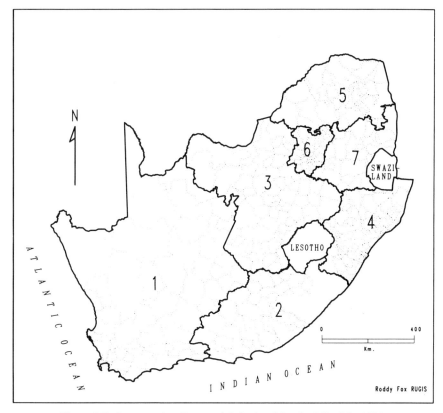

Figure 2.7. Seven-region Proposal Submitted by the NP, July 1993

there is scant support for President Mangope in these two districts (Drummond, 1991).

One point that needs to be elucidated applies to both the National Party and CDDR proposals. Both of them justify their regional boundaries using different criteria at different times. This inevitably leads to inconsistencies and the thought that particular interests were being served and justified, using whichever criterion appeared appropriate. This has also made interpreting the demarcations a perplexing exercise.

THE CDDR'S PROPOSAL

Figure 2.8, the CDDR's proposed map (Commission on the Demarcation/Delimitation of SPRs, 1993a) can be interpreted as a simple compromise between the ANC and National Party proposals. The National Party gained acceptance for a

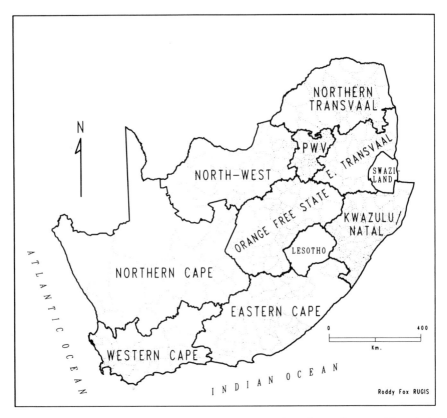

Figure 2.8. Original Demarcation Commission Proposal to the Multi-party Negotiating Council

separate Western and Northern Cape, the ANC's proposed Eastern Cape was delimited and the Orange Free State and North-west remained separate. Dissensions within the Committee on these boundaries varied from region to region and were reported as: Northern Transvaal, one dissension; PWV, two dissensions; Eastern Transvaal, one; KwaZulu/Natal, none; Orange Free State, one; North-west, one; Northern Cape, seven; Western Cape, seven; Eastern Cape, one. Quite clearly the principal problem area was the delimiting of a separate Western and Northern Cape. The other problem areas identified by Ms Bernstein, one of the two minority dissenters who refused to sign the report were: the number of regions in the Eastern Cape; black opinion on the Northern Transvaal region; the effect of a separate Northern Cape on the poor in that region; the debate over the separation of the Pretoria functional area from the greater PWV.

The CDDR report suffered from many of the problems already listed previously. In addition it drew criticism from two of the Commissioners: from the ideological right, Mr Koos Reyneke; and the left, Ms Anne Bernstein. Mr Reyneke refused to sign the report because it rejected the possibility of a Volkstaat and because of the lack of 'conflict reducing socio-cultural borders' (Commission on the Demarcation/Delimitation of SPRs, 1993a, p. 82). Ms Bernstein had different problems: the very short time span with which the Commission was faced; the undemocratic nature of the process which had not drawn responses from the marginalized peripheral communities (echoing comments made in the ANC submission); and the roles adopted by the Commissioners who took it upon themselves to negotiate the compromises rather than present the politicians with alternatives and allow them to negotiate. The Negotiating Council of the Multi-Party Negotiating Process accepted the report as a useful starting point but recommended that oral and written evidence was necessary in respect of the sensitive and contentious areas already identified by the CDDR. Thus they were in part responding to criticisms by Ms Bernstein and the ANC. Furthermore, the second CDDR report does not attempt to arrive at a delimitation but to present arguments, for and against, in an objective manner for each contentious area in turn. In this way they were also responding to Ms Bernstein's criticism of the role that the Commissioners had played in the first delimitation exercise.

THE CDDR'S SECOND REPORT AND FINAL BOUNDARY RECOMMENDATIONS

The CDDR's second report (Commission on the Demarcation/Delimitation of SPRs, 1993b) shows that once again a very large number of submissions were forthcoming in a very short space of time. In this case there were 467 written and 177 oral submissions. Between 11 September and the presentation of the second report on 15 October these were to be assimilated, synthesized and presented as cases for and against each contentious area and then related to the boundary criteria.

Many sensitive areas were examined; most of them are indicated in Figure 2.9, which was compiled from the final boundary recommendations arising from the submission of the second report. Figure 2.9 shows them as areas where district adjustments were accepted, areas where referendums could take place and boundaries between regions which might fall away entirely. The issues considered were as follows:

- Eastern Transvaal, the positioning on its north and north-western borders of Pretoria, KwaNdebele, Bronkhorstspruit, Middelburg and Witbank, the sharing of Kruger National Park with Northern Transvaal.
- Northern Transvaal, the positioning of Pretoria, Groblersdal, Pilgrims Rest and Hammanskraal on its southern border and the sharing of Kruger National Park;

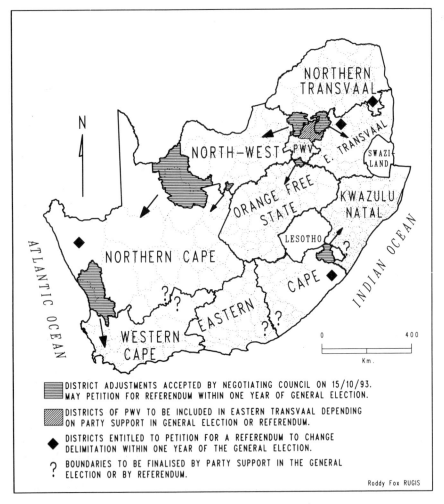

Figure 2.9. Final Boundary Adjustments Adopted by the Negotiating Council, November 1993

- PWV, whether this region should exclude Sasolburg, Pretoria and the Odi and Moretele portions of Bophuthatswana;
- Orange Free State, whether this region be combined with the North-West;
- KwaZulu/Natal, the inclusion on its southern border of Umzimkulu and Mount Currie;
- Northern Cape, whether it should include Kuruman and Postmasburg on its northern border but exclude Namaqualand on its southern border;

- Eastern Cape, whether this region should be split into two, the Eastern Cape and Border/Kei;
- Western Cape, whether this region be combined with the Northern Cape.

The second report acknowledges that the Commission experienced a number of difficulties: delimiting regions without knowing their powers; the status of the Kruger Park; the relative importance of the delimiting criteria. For example, they could not decide on the location of Umzimkulu, Xhosa-speaking and administratively part of Transkei but functionally in the hinterland of Durban. The same problem occurred with districts along the Eastern Transvaal, Northern Transvaal border. In another case, Pretoria's incorporation in Eastern or Northern Transvaal would undoubtedly strengthen the economic resource base of these regions but would then cut across the functional linkages Pretoria had with the rest of the PWV. Another interesting feature of the report was the prominence now being given by interested parties to the concept of sub-regions.

The Negotiating Council resolved the issues presented to them through coordinating a sequence of bilateral meetings between the interested political parties (Commission on the Demarcation/Delimitation of SPRs, 1993c). Referendums were proposed by the bilateral meetings as an option for resolving disputed areas, but there was no political consensus as to whether these could be organized by February 1994, only two months before the 27 April 1994 elections. Consequently referendums were proposed as an option which disputed areas could take up after the general election.

The Council adopted a number of changes to the regional boundaries, shown as shaded areas in Figure 2.9, subject to the referendum proviso. Referendums could also be called by districts whose delimitation has not been changed, for example Namaqualand in the Northern Cape. Finally, the major boundary problems between the Northern and Western Cape, between the Eastern Cape, Border/Kei and Umzimkulu, and the positioning of the greater Pretoria area were to be settled by a further mechanism. The votes in the election were to be counted separately in these regions and then amalgamated by party support to decide the issue 'subject to a majority agreement between the parties elected by the voters of the area, or subject to a referendum' (Commission on the Demarcation/Delimitation of SPRs, 1993c, p. 2). In the Eastern Cape and Northern Cape a majority was defined as 60 per cent, in Pretoria it was 50 per cent.

The way this would work can be seen from the following example. In the Eastern Cape, if there was more than 60 per cent support for the National Party and Democratic Party in the portion of the region west of the Fish River dividing line, then that portion would become a separate region. Similarly, if the Pan Africanist Congress (PAC) received more than 60 per cent of the votes in the portion east of the Fish River, then that would become a separate region. The NP, DP and PAC were clearly supporting separate regions for the areas in which they perceived the majority of their supporters to lie.

CONCLUSIONS

This chapter has attempted to follow and assess the geographical aspects of the debate over regions for the 'new' South Africa; a regional demarcation process which will continue through the activities of the Commission on Provincial Government. Key influences on the minority parties have been the spatial distribution of the electorate, particularly their ethnic and racial components, and thus the potential for alliances to amalgamate regional voting blocs. The relative wealth of the country's regions has been another influence on all of the major players. The final demarcation process itself was hurried, although a surprisingly large number of submissions were examined, and inevitably flawed as there was little clarity over what the principal delimiting criterion should be. The process was also hampered by the delinking of the boundaries of regions from their powers in the hope that this could produce a technically neutral solution to a political problem. Perhaps it is appropriate to conclude with a quotation from the ANC's submission to the CDDR (African National Congress, 1993, p. 14) where they complete their discussion of the principles utilized internationally to delimit regions: 'Delimitation is a fundamentally political process. There is no technically correct region, based on criteria—regions are manufactured politically and then rationalized later.'

REFERENCES

African National Congress (1993) *The Delimitation of Regional Boundaries*, submission 107, Commission on the Demarcation/Delimitation of SPRs, 31 July.

Beeld (1992) *Nuwe gedagtes oor streekregerings, Beeld*, Johannesburg, 15 April, p. 15.

Commission on the Demarcation/Delimitation of SPRs (1993a) *Report of the Commission on the Demarcation/Delimitation of SPRs, 31 July 1993*, Multi-Party Negotiating Process, Johannesburg.

Commission on the Demarcation/Delimitation of SPRs (1993b) *Report on Further Work on the Demarcation/Delimitation of States/Provinces/Regions (SPRs), 15 October 1993*, Multi-Party Negotiating Process, Johannesburg.

Commission on the Demarcation/Delimitation of SPRs (1993c) *Report by the Co-ordinating Committee on the Demarcation/Delimitation of SPRs as Adopted by the Negotiating Council on 15 November 1993*, Multi-Party Negotiating Process, Johannesburg.

Constitutional Committee of the African National Congress (1992) *Ten Proposed Regions for a United S.A.*, Centre for Development Studies, University of the Western Cape, Cape Town.

Daily Dispatch (1992) 'ANC outlines plan for transitional government', East London, 9 November, p. 9.

de Kock, C., Mareka, C., Rhoodie, N. and Schutte, C. (1993) *The Prospects for a Free, Democratic Election: Inhibiting and Facilitating Factors in Voting Intention*, Human Sciences Research Council, Pretoria.

Democratic Party (1992) *Discussion Paper on Constitutional Proposals*, National Policy Advisory Committee, Democratic Party, Cape Town.

Development Action Group (1992) *Comments on the Government's Seven Region Proposal*, unpublished paper, Development Action Group, Cape Town.

Drummond, J. (1991) 'Reincorporating the bantustans into South Africa: the question of Bophuthatswana', *Geography*, 76(4), 369–73.

Eastern Province Herald (1992) 'X-factor will be decisive on the day', Port Elizabeth, 28 September, p. 6.

Fox, R. C. (1992) 'A piece of the jigsaw: the 1992 white referendum and South Africa's post-apartheid political geography', *Journal of Contemporary African Studies*, 10(2), 84–96.

Fox, R. C. (1994) 'South Africa: the legacy of apartheid' in T. Unwin (ed.), *Atlas of World Development*, John Wiley, Chichester, 314–316.

Grobler, E., Prinsloo, K.P. and van der Merwe, I.J. (eds) (1990) *Language Atlas of South Africa: Language and Literacy Patterns*, Human Sciences Research Council, Pretoria.

Humphries, R. (1992) 'National Party and state perspectives on regionalism', *Africa Insight*, 22(1), 57–65.

Humphries, R. and Shubane, K. (1992) 'A delicate balance: reconstructing regionalism in South Africa', *Centre for Policy Studies, Transition Series Report 24*, Johannesburg.

McCarthy, J. (1993) 'Factual context to political competition', *Development and Democracy*, 5 (July), 39–49.

Multi-Party Negotiating Process (1993) *Document Pack for the Meeting of the Plenary of the Multi-Party Negotiating Process, 17 November 1993, Volume IV*, Multi-Party Negotiating Process, Johannesburg.

National Party (1993) *Proposals by the Government of the Republic of South Africa on Boundaries for Regions to Establish Regional Government in South Africa*, Submission 117, Commission on the Demarcation/Delimitation of SPRs, 31 July.

Pienaar, P.A. (1992) *A Model for the Delimitation of Autonomous Regions in South Africa*, Department of Geography, University of South Africa, Pretoria.

Republic of South Africa (1988) 'Regional development in Southern Africa and an exposition of the composition, aim and functions of the NDRAC, RDACs, RDAs, and DDAs', *Supplement to Informa*, April, Department of Development Planning, Pretoria.

Republic of South Africa (1992) *Report of the Committee for Constitutional Affairs on a Proportional Polling System for South Africa in a New Constitutional Dispensation*. PC 2/1992, Government Printer, Cape Town.

Sunday Times (1994) 'ANC is heading for April 27 landslide', Times Media Ltd, Johannesburg, 16 January.

Weekly Mail (1992) 'Revealed: ANC election strategy', Johannesburg, 9–15 October, p. 5.

Wessels, E. (ed.) (1993) 'Evil empire ends at midnight', *Negotiation News*, 13 (15 November), 6–11.

Part II

DOMESTIC POLICY QUESTIONS FOR THE NEW GOVERNMENT

3 Redistribution and Social Justice after Apartheid

DAVID M. SMITH

It is sometimes supposed that the abolition of apartheid legislation, followed by the implementation of a non-racial democracy, will deliver social justice to South Africa. This chapter questions such an assumption. It brings together some theoretical deliberations on social justice and some facts concerning the racial inequalities generated by apartheid to explore the possibilities for redistribution and what this means for social justice after apartheid.

The theoretical deliberations arise from a renewed interest in social justice in geography. While social justice has never been far below the surface of concerns with such issues as race and gender, there are signs of a return to the more explicit treatment which formed part of the so-called radical geography of the 1970s (see Smith, 1994). That social justice has continued to interest moral and political philosophers is indicated by the volume and scope of the literature accumulated in the past two decades (see, for example Kymlicka, 1990). In a world of rapid change, not least in South Africa, the resurrection of social justice in geography could hardly be more timely.

In the first volume of his contemporary treatise on social justice, Brian Barry (1989, p. 3) reminds us:

> social arrangements are not a natural phenomenon but a human creation. And what was made by human beings can be changed by human beings. This realization sets the stage for the emergence of theories of social justice. For a theory of social justice is a theory about the kind of social arrangements that can be defended.
>
> In Plato's time as in ours, the central issue in any theory of social justice is the defensibility of unequal relations between people. Like the Athenians, we see all around us in our societies huge inequalities in political power, in social standing, and in the command over economic resources. The degree of inequality on each of these dimensions is different in different societies, and so is the extent to which a high position on one is associated with a high position on the others. South Africa is not easily confused with Scandinavia.

That South Africa is distinctive is hard to dispute. So is the proposition that apartheid society was (and its relics remain) grossly unjust. But it is more difficult to say what kind of post-apartheid society, incorporating what kind of inequalities, might most persuasively be defended on grounds of social justice. A central issue in any attempt to apply theoretical thinking about social justice to

The Geography of Change in South Africa, edited by Anthony Lemon.

the particularity of one country is the extent to which the prescription of a just society can be guided by some broad or even universal principle(s), as opposed to the relativist position that what is socially just is historically and locally specific, or contextual. The argument underpinning the discussion which follows is that a form of egalitarianism provides an appropriate universal framework to guide thinking about social justice in South Africa but that its application requires attention to the particular inheritance of apartheid.

It should be added that the analysis here is confined to the Republic of South Africa, as defined before the independence of certain so-called 'homelands' began with the Transkei in 1976. No consideration is given to the implications of economic, political and social change on neighbouring states, including land-locked Lesotho, the fortunes of which are closely bound to those of South Africa. Ultimately, of course, the question of whether the world is better (or more socially just) for the demise of apartheid cannot be resolved within South Africa's boundaries alone.

DISTRIBUTION AND SOCIAL JUSTICE

While social justice is a very broad concept, attention is usually focused on the unequal distribution of income and other sources of need satisfaction on which the material conditions of a population depend. To echo Brian Barry, it is *in*equality that requires justification. People's common humanity and capacity for pleasure and pain, along with the same basic needs, is a plausible starting point for an egalitarian argument, with such individual differences as strength, skill, intellect, family or place of birth regarded as fortuitous and hence morally irrelevant to how people should be treated. Racial identity or classification is, of course, one of these differences over which people have no control. That racial discrimination is morally indefensible is thus a widely held view, with racial inequalities in life chances indicative of social injustice.

However, there are grounds on which different and unequal treatment of people can be justified. Some people may be deemed to deserve more or less of what there is to distribute, for example if they produce more than others or occupy positions of special responsibility. Greater or lesser contribution to the common good is often held to justify unequal rewards according to the quantity or quality of people's work. Need is another common criterion for unequal treatment. Some societies have also been able to construct and maintain ideologies justifying the favourable (or otherwise) treatment of some kinds of people according to such morally irrelevant criteria as family origin, gender or race; such was the case in South Africa under apartheid.

A possible point of theoretical departure is to try to identify an initial state of affairs which can be accepted as just and to argue that any outcome will also be just providing that it arises from a just process. An example can be found in the libertarianism of Robert Nozick (1974). If peoples' holdings of property (such as land and natural resources) have been justly acquired, for example by settling land held by no one else or purchasing it by mutual agreement, and if they subsequently use this property justly to acquire further wealth, for example by free trade or mutually agreed employment of others, then the distributional consequences can be justified no matter how unequal. There is no particular pattern to which a just distribution should conform. Such an argument is sometimes used to justify the distribution generated by market forces, assumed to embody a just process, though the justice of the prior distribution (including how people actually acquired their property in the first place) may not be closely scrutinized.

This perspective would not take us very far in South Africa, given the obvious injustice of the distribution of property from the earliest days of European settlement. Not even the most ardent libertarian would accept plunder, theft and coercion as a just basis for the initial distribution, even if a just set of institutions had subsequently prevailed (which was clearly not the case). Similarly today, the existing distribution of property could not possibly comprise a basis whereby a new and just process of (re)distribution could generate a just outcome.

An alternative perspective is to be found in social contract theory, the best known contribution to which is the work of John Rawls (1971). Although subject to extensive subsequent critique, Rawls's theory of justice is still extremely influential. The approach adopted by Rawls was to try to deduce the social contract to which people would subscribe in particular circumstances. He began with an 'original position', in which people would have to approve of institutions under a 'veil of ignorance' as to their actual situation, for example whether they would belong to the rich or poor. Applied to South Africa, people would have to agree a new constitution, not knowing the race group to which they would belong (or how they would have been classified under the Population Registration Act); or not knowing whether they would end up living in Ilovo or Soweto, Crossroads or Constantia, for example. Not knowing who, and where, they would be in society, most people would be prepared to entertain only a narrow range of life chances, if any inequality at all, because they could end up among the worst-off (though some might be prepared to take the risk of ending up very poor for the sake of the chance of being very rich).

Rawls (1971, p. 303) states his general conception of justice as follows:

> All social primary goods—liberty and opportunity, income and wealth, and the bases of self-respect—are to be distributed equally unless an unequal distribution of any or all of these goods is to the advantage of the least favoured.

This now famous dictum is known as the 'difference' or 'maximin' principle because it requires the maximization of the conditions of those at the minimum level. Barry (1989, pp. 225–6) argues that this principle does not actually depend on Rawls's unrealistic original position: all that is required is to accept that everything about the sources of occupational achievement on which inequality is based are contingent and morally arbitrary—which requires the equal distribution of social primary goods—but that some inequality can nevertheless be justified on the grounds that it benefits all and especially the worst-off.

Rawls (1971, p. 302) elaborates his principles of justice as follows:

First Principle
Each person is to have an equal right to the most extensive total system of equal basic liberties compatible with a similar system of liberty to all.

Second Principle
Social and economic inequalities are to be arranged so that they are both:
(a) to the greatest benefit of the least advantaged, consistent with the just savings principle [required to respect the claims of future generations], and
(b) attached to offices and positions open to all under conditions of fair equality of opportunity.

Application to South Africa would require that whites had acquired their positions of advantage in a situation of equal liberties and opportunities for all people irrespective of race and that white privilege was to the advantage of the poorest blacks.

An elaboration of Rawls's original theory has been provided by Rodney Peffer (1990), writing from a Marxian perspective. He departs from the position, common among Marxists, that morality and social justice are ideological constructs deployed by a ruling class to legitimate their position of privilege, to recover what he sees as the moral theory implicit (for the most part) in Marx's writings. He concludes with a modification of Rawls's theory (see Peffer, 1990, p. 418, for a statement of principles). The main effect is to prioritize peoples' rights to security and subsistence, to make more specific certain liberties and opportunities which should prevail (including participation in all social decision-making processes) and to require inequalities not to exceed levels which would seriously undermine liberty and self-respect. While the superiority of Peffer's position over Rawls's original formulation is questionable (see Smith, 1994, Chapter 4), it does point to some congruence between liberal and Marxian positions on distributive justice. The major difference, of course, is that whereas Rawls and like minds see social justice associated with liberal, democratic institutions, Marxists see socialism as the essential prerequisite and revolutionary change a legitimate means of achieving societal transformation. Even if socialism is rejected in favour of capitalism, however, application of Rawls's difference principle has profound egalitarian implications (see Miller, 1992).

If we accept as implausible any suggestion that the existing inequalities in South Africa are in the interests of the worst-off or an outcome of equal opportunities to acquire positions of advantage, it follows that moves towards racial equality will improve social justice even if perfect equality is not achieved. This position is also consistent with the moral irrelevancy of race as a distributive criterion. The only other consideration that should be added, before turning to South Africa, is that of reparation, or compensation for past injustice, which might strengthen an egalitarian position, influence the design of strategy for its implementation, and support such measures as affirmative action and positive discrimination.

RACIAL INEQUALITY UNDER APARTHEID

The most obvious expression of inequality is the distribution of income by race group. One attempt to identify this is reproduced in Table 3.1. The top part shows that in 1987 whites earned 62 per cent of South Africa's income (for only 14 per cent of the population), whereas blacks accounted for only 27 per cent of the income (for almost 75 per cent of the population). More recent data are as follows: 39.4 million blacks comprising almost 76 per cent of South Africa's population receiving 33 per cent of personal income, 5.1 million whites (almost 13 per cent) with 54 per cent of income, 3.4 million coloureds (8.6 per cent) with 9 per cent of income, 1.0 million Indians (2.5 per cent) with 4 per cent of income (1992 populations from SAIRR, 1993, p. 254; 1990 income data from SAIRR, 1992, p. 257, after Charles Simkins).

Table 3.2 provides figures for a selection of economic and social indicators by race group. The ratio of advantage of whites over blacks (W:B) varies from over 14 for the tuberculosis rate to only just over one for life expectation and for old age pensions (where equality has subsequently been achieved). Household

Table 3.1. Distribution of Income by Race Group, 1936–87

Race group	1936	1960	1980	1987
Personal income (per cent)				
Africans/Blacks	18.7	19.9	24.9	27.0
Asians/Indians	1.7	2.1	3.0	3.5
Coloureds	4.2	5.6	7.2	7.5
Whites	74.9	72.5	64.9	62.0
Per capita income (R)				
Africans/Blacks	460	795	1284	1246
Asians/Indians	1395	1677	3864	4560
Coloureds	942	1563	2900	3000
Whites	6033	9810	15 180	14 880
Whites: Africans/Blacks	13.12	12.34	11.82	11.94

Source: SAIRR (1989, p. 423), after S. J. Terreblanche. The figures refer to the whole of South Africa, including the 'independent' homelands. The personal income percentages for 1936 and 1960 incorporate arithmetic errors in the source.

Table 3.2. Selected Economic and Social Indicators by Race Group

Indicator	Black	White	Coloured	Indian	W:B
Average monthly household income (R) 1991	779	4679	1607	2476	6.01
Average monthly earnings in manufacturing (R) 1991	1056	3793	1187	1727	3.59
Monthly old-age pensions (R) 1992	293	345	318	318	1.18
Infant deaths per 1000 live births 1990	52.8	7.3	28.0	13.5	7.23
Life expectation at birth (years) 1987	63	73	63	67	1.16
Tuberculosis (new cases per 100 000 population) 1987	216.7	14.8	579.8	53.2	14.64
Expenditure on school pupils per capita (R) 1991–2	1248	4448	2701	3500	3.56
Pass rate in matriculation examinations (%) 1991	41	96	83	95	2.34

R = Rand; R1.00 = approx £0.20.
Source: SAIRR (1993). The figures generally refer to the Republic of South Africa, without the 'independent' homelands.

incomes and infant mortality reveal disparities at about six or seven to one, although the official unemployment figures here are likely to be a substantial undercount. Earnings in manufacturing and expenditure on school pupils vary at around 3.5:1. Coloureds and Asians occupy positions between blacks and whites, the Asians generally doing better.

An obvious indication of the pace of equalization now required is provided by what has actually been taking place, for post-apartheid society must surely improve on this. Table 3.1 shows that there has been some increase over the years in the share of income going to blacks, and also to coloureds and Asians, but this has been partially offset by more rapid population growth than for whites. The coefficient of concentration, which compares the percentage distribution of income with population among the four groups (on a scale of 0 to 100, where 0 indicates exact correspondence or perfect equality and 100 the extreme of inequality), gives 59.5 for 1936 and 48.9 for 1987. The Gini coefficient of inequality calculated on the same data registers less of a decrease, from 54.9 in 1936 to 53.5 in 1987. The bottom part of Table 3.1 shows how per capita income has changed: the ratio of white to black income indicates a narrowing of the gap from 1936 to 1980, but there is a slight reversal in the 1987 figure.

Figures for more recent years show the share of personal incomes going to blacks at 29 per cent in 1985, 33 in 1990 and 37 in 1995 (estimated, assuming the optimistic annual growth rate of 2.5 per cent); the white shares in these same years are 59, 54 and 49 per cent (SAIRR, 1992, pp. 256–7, after Charles Simkins). Changes of these magnitudes would take 40 to 50 years for race-group shares of income to approximate shares of national population.

Other indications are provided in Table 3.3 by figures for earnings in manufacturing over the 1970s and 1980s. The ratios of advantage of whites over the other race groups indicate steady convergence. However, projecting the average rates of change over this period into the future, it would take about 20 years for black and white earnings to reach parity and the same for Indians and whites; for coloureds and whites it is about 50 years because of the relatively slow rate of improvement in coloured earnings. Changing disparities in wages in the construction industry are of similar magnitudes: the ratios of advantage between whites and blacks has been reduced from 6.7 in 1970 to 4.5 in 1990, and between whites and Indians from 2.3 to 2.0, while between whites and coloureds there was an increase from 2.9 to 3.1 (McGrath and Holden, 1992, p. 36).

Two other sets of indicators of the pace of equalization may be summarized briefly. The number of years of life expectation at birth increased between 1950 and 1987 from 45 to 63 for Africans, 46 to 63 for coloureds, 55 to 67 for Indians and 67 to 73 for whites (SAIRR, 1993, p. 280). State education expenditure per capita on black children rose between 1969/70 and 1989/90 from 5 per cent of the white figure to 25 per cent, for coloureds from 20 per cent to 53 per cent, and for Indians from 27 to 72 per cent (SAIRR, 1992, p. 195).

It is obvious from these figures that, although inequality among the race groups, on a range of conditions, has been narrowing under apartheid, the pace of equalization will have to be greatly accelerated if anything approaching racial equality is to be achieved before well into the next century.

Table 3.3. Average Annual Earnings in Manufacturing (Rand) by Race Group, 1972–89

Year	White	Black	Coloured	Indian	W:B	W:C	W:I
1972	4308	782	1080	1632	5.51	3.99	3.63
1975	6132	1272	1620	1812	4.82	3.79	3.88
1980	11 472	2688	3156	3588	4.27	3.63	3.20
1985	22 188	5628	6372	7980	3.94	3.48	2.78
1989	38 334	11 105	11 682	15 682	3.45	3.28	2.44

Source: SAIRR (1992, p. 259).

REDISTRIBUTION: OPPORTUNITIES AND CONSTRAINTS

There are three possible redistributional strategies to accelerate equalization of living standards. The first is reallocation of public expenditure to enhance the capabilities of the poor to increase their earnings and general well-being within the existing economic structure. The second is redistribution of wealth, income and other assets through taxation for example, so as more directly to relieve hardship and increase the consumption possibilities of the poor. The third is structural change, for example in ownership or control of the means of production and in the distribution of economic and political power, so as to alter or eliminate the basic mechanisms generating inequality.

We begin with the first and least radical of these strategies. Once the end of apartheid has established formal equality of opportunities in all spheres of life, the next step is to ensure that everyone has the same capacity to take advantage of opportunities. As Eckert (1991, p. 45) explains, enormous differentials in capabilities exist today as the result of apartheid and earlier practices of discrimination: 'Simply ensuring an equal chance in the future is not enough when the capabilities of so many have been artificially lowered by past practice.'

Levels of service provision have a bearing on capabilities, and equal treatment is the least that might be expected. Indeed, there is a case for positive discrimination in favour of those whose greater need is manifest in lower living standards. To improve social services for the black population is an obvious priority, but the task of actually achieving this is formidable. Van der Berg (1991, pp. 76–7; 1992, pp. 130–2) has shown that to equalize up to prevailing white levels of expenditure on primary and secondary education, health and social pensions would require an inconceivable threefold increase in total spending: see Table 3.4. To equalize within the present budget constraint would require expenditure on whites to be cut to about one-third of current levels with that on coloureds and Indians also reduced. To equalize educational expenditure alone at white levels in 1987 would have required expenditure to have been trebled (Hofmeyr and McLennon, 1992, p. 184). Another calculation suggests that racial equalization of social spending, including housing, at white levels would increase this budget item from 9.5 per cent of GDP (a figure not out of line for a country with South Africa's GDP per capita) to 30.8 per cent (Moll, 1991, p. 125; van der Berg, 1992, p. 135). At the root of the problem is the fact that, under apartheid, spending on whites was raised to levels unsustainable for the entire population in a normal country at South Africa's overall level of economic development.

Resources required for equalizing expenditure on services can be found in three ways. The first is economic growth, the most comfortable method because gains to the poor need not be at the expense of the rich. However, the past decade has seen South Africa's annual growth in GDP average less than 1.0 per cent, compared with population growth of about 2.5 per cent, with a consequent reduction in GNP per capita from R3981 in 1980 to R3060 in 1993 (in constant

Table 3.4. Actual and Parity Requirements in Government Social Spending, 1986–7

Category	White	Coloured	Indian	African	Total
Total social spending (R bn)					
Actual	4.3	1.7	0.5	5.4	11.9
Parity: at white levels	4.3	3.5	0.9	25.5	34.2
within current constraints	1.5	1.2	0.3	8.9	11.9
Population (1000s)	4892	3013	915	25 158	33 978
Per capita social spending (R)					
Actual	879	654	547	214	361
Parity: at white levels	879	1162	985	1013	1036
within current constraints	307	398	323	353	361

Source: van der Berg (1992, p. 132). Social spending includes primary and secondary education, health and social pensions. Note that the differences between the groups in per capita spending required for parity arise from differences in age structure determining clientele for education and pensions and from differences in income causing fewer whites to require social pensions.

prices; from *Financial Times*, 11 June 1993). The second way is by transferring resources from other groups to make up the shortfall in expenditure on blacks. However compelling the moral justification for such redistribution may be, white vested interests tend to focus attention on a third strategy, that of reallocation of expenditure between sectors within the existing state budget.

Abandoning the formal apparatus of apartheid should facilitate some reallocation of public expenditure. Van der Berg (1991, p. 80) estimates that the elimination of institutions (for example, 14 separate departments of education, covering South Africa's four official race groups and the ten bantustans), could save about 1 per cent of GDP. Rather more could be found by reducing defence expenditure: if this item had not risen above its 1972 level of 2.2 per cent of GDP the subsequent saving could have built 1.9 million fully serviced houses which is well over the estimated national shortage (van der Berg, 1991, p. 81). Moll (1991, pp. 122–5) suggests that a reduction in defence expenditure by 2 per cent of GDP, the abolition of apartheid institutions and the elimination of the industrial decentralization programme, along with a 'modest' increase in taxes and government borrowing on a 'prudent' scale, could yield the equivalent of 6.4 per cent of GDP. This compares with 21.3 per cent needed to equalize social spending at white levels. He concludes that equalization at current white levels is impossible at present, given South Africa's level of economic development, and will have to be achieved at substantially lower levels of expenditure per capita.

Reallocation of priorities within particular services provides some redistributive opportunities. In education (see Chapter 7), the obvious priority is African primary education to address massive underachievement as well as provide for large numbers of children not even in school. But past inequalities are reflected in fixed capital as well as pupil–teacher ratios; it will take reallocation of premises and personnel as well as current expenditure to implement effective redistribution. The location of existing services is also an impediment to redistribution in health care (see Chapter 6), where the priority is to shift resources to basic primary care, especially in the informal settlements and rural areas, and away from high-cost intervention which favours well-to-do (mainly white) city dwellers.

More state expenditure is required in skills training and employment generation (see Chapter 9) in an economy where less than one-tenth of new labour market entrants now find work in the formal sector. Participation in the informal sector is difficult to measure, but an estimate for 1991 by the Development Bank for Southern Africa put the figure at four million or almost 30 per cent of the economically active population (SAIRR, 1992, p. 173). Reallocation of funds currently benefiting big business, to stimulate further the absorption capacity of informal activity, is an obvious strategy. The encouragement of small-scale 'appropriate technology' approaches to the production of 'affordable' housing (see Chapter 8) could create jobs as well as much needed shelter in the cities.

The severe constraints imposed on reallocation of public expenditure lead to the second, more direct strategy for redistribution. Increasing personal taxation is constrained by the fact that South Africans are already relatively highly taxed by international standards. However, the burden is mainly on middle-income earners, and increasing taxes on high personal incomes, and on businesses via capital gains and capital transfer taxes, are realistic possibilities (Davis, 1992). A general wealth tax should not be ruled out.

However, such measures face strategic difficulties, as well as opposition from those likely to lose. Most whites could sustain some losses without great discomfort, but there are knock-on effects to consider. Tax increases on businesses may induce capital flight and the loss of valuable personnel, while tax increases on individuals could put the livelihood of many of the hundreds of thousands of (black) domestic servants at risk. And there are also considerations of employment in industries and services currently catering to the demands of affluent whites, if their spending power were to be reduced.

A recognition that measures which are supposed to promote positive redistribution can have undesirable outcomes can encourage undue caution, especially when coupled with white self-interest. The 'realism' prevailing in establishment circles needs to be leavened by more imaginative thinking. For example, distributing free shares to gold-mining workers could be an effective way of allowing the market to redistribute the value of company assets from current shareholders to workers. Redistribution of the existing housing stock may be inconceivable politically, but property taxes could encourage small white families in large homes to vacate them, making it easier for larger black households to find accommodation suited to their needs. Black township housing could be given away to existing tenants who may have already effectively paid for it in rent.

Privatization of some state corporations could be accomplished by giving shares to all South Africans on an equal basis (Reekie, 1990, pp. 13–14) along the lines of privatization strategies in some post-socialist countries such as Russia where vouchers for shares have been distributed in this way.

The third strategy for redistribution is structural change. Up to the time of Nelson Mandela's release in 1991, the widespread belief of the liberation movement was that a socialist transformation would ensure rapid redistribution in a post-apartheid society. But the collapse of socialism in Eastern Europe and the former Soviet Union has led to a growing recognition that South Africa is likely to remain capitalist. This is not to say that such measures as public ownership are now completely off the agenda, however. Nationalization was a central feature of the Freedom Charter, and the possibility of state appropriation of the mining houses and banks continues to be raised. But this increasingly seems a reminder of what could happen if big business is insensitive to the needs of a new society rather than an indication of serious intent.

It is also questionable what nationalization would actually achieve, with respect to the redistribution of wealth. Moll (1990, pp. 78–9) has estimated that wholesale nationalization well beyond the 'commanding heights' mentioned in the Freedom Charter would enable the state to redistribute about R900 a year to the poorest 40 per cent of the population (compared with annual incomes of about R1400 for black rural dwellers and R9200 for mining workers). But the process of disinvestment that nationalization would provoke, and the consequent negative impact on economic growth, could make this kind of redistribution merely a one-off bonus to the poor. The fact is that with the pace of redistribution dependent on stimulating the economy, international capital has the power to constrain a new South African government's freedom to change existing patterns of property ownership. Thus sustainable redistribution via large-scale public take-over of private business no longer seems to be a realistic proposition.

However, there are possibilities for structural adjustments to the economy short of revolutionary transformation. The slow rate of employment generation, which is a root cause of poverty, can be attributed in part to private investor preference for property speculation at the expense of the manufacturing sector. The reallocation of investment in directions more consistent with basic human needs is a favoured strategy in progressive circles; the ANC stresses 'growth through redistribution' to stimulate development more responsive to the interests of the mass of the people. The power already exercised by ordinary people to influence business behaviour, for example in disinvestment and boycott campaigns, may be transferable to other arenas of struggle with the business establishment (Bond, 1991). A new government could encourage an environment in which greater social responsibility by local business becomes part of the broader democratization of South African society. However, such a strategy would require sensitivity to the reaction of foreign companies, who sometimes saw themselves in the vanguard of enlightened practices under apartheid, if new investment is not to be discouraged.

It will be evident from this review of the possibilities for redistribution that the prospects for rapid equalization in post-apartheid society are not encouraging. The constraints on redistribution are severe, and there is a danger, even in the more radical strategies, that transfers of resources from the relatively rich may not flow right down the socio-economic scale. For example, transferring assets of corporations to their work-force benefits those who are already relatively well paid by the standards of South African blacks, leaving out millions who currently have no prospect of formal-sector employment. This kind of redistribution could also be highly selective from a geographical point of view, benefiting mainly the residents of formal urban areas in the major metropolitan regions, without reaching far into either the peripheral shack settlements or the rural areas where the poorest people live. In short, vast numbers of the worst-off might stand to gain very little, which is hardly a prescription for social stability let alone social justice.

RESOURCES, REGIONS AND CITIES

The redistribution of land and other natural resources raises different questions from those considered above (see Chapter 4 and Smith, 1994, Chapter 8 for a fuller discussion). Social justice does not require the equal distribution of land, for example; the question is that of who is entitled to particular parcels for particular uses. That this is a matter of great contention is obvious, given the claims arising from various forms of dispossession which deprived blacks of farmland and Indians and coloureds as well as blacks of urban residential sites. White appropriation of land held by other people was evidently unjust, but returning property to its original holders is far from straightforward. There are practical problems in establishing who should ultimately be entitled to particular parcels, and also legitimate rights of subsequent owners to enhanced land values for which their labour or financial investment was responsible.

So it is not a simple case of turning the clock back, to reinstate some original pattern of just holdings. Circumstances change, and it is important to consider the just treatment of present people as well of those wronged in the past, as Waldron (1992) stresses in a discussion of historic injustice. Nor is it satisfactory simply to let bygones be bygones. This is, in effect, what was proposed in the *White Paper on Land Reform* (South Africa, 1991), which sought to legitimate land holdings arrived at under apartheid (and earlier forms of racial domination), with subsequent exchange achieved through a supposedly deracialized land market responding to entrenched racial inequalities in purchasing power (see Marcus, 1991, for further discussion). Land reform should contain an element of restitution to past holders, with a land court set up to resolve conflicts (Claassens,

1991). It should also involve acquisition of additional land for black farmers previously dispossessed or otherwise deprived of the opportunity to make a living from agriculture. A land value tax could provide revenue for this purpose (Moore, 1992; Davis, 1992). A tax on urban land values also provides a way for existing owners—many of whom may have gained from Group Areas removals—to compensate the dispossessed (or their descendants) without the disruption of large-scale restitution of residential property. The practical obstacles to such measures are probably no more severe than those faced by the return of property to original owners in some post-socialist countries in East Europe, such as Poland.

Uneven regional development is understandably eclipsed by racial inequality. Yet region of residence is as morally arbitrary as race with respect to the differentiation of life chances. The extent of disparities in Gross Geographical Product (GGP) among the nine official development regions (see Fig. 2.1) is indicated in Table 3.5. The per capita index for 1987 shows that the region with the highest GGP in relation to population is H (Pretoria–Witwatersrand–Vereeniging) and the lowest D (containing the Transkei and Ciskei). The enormous gulf between the metropolitan areas and homelands is shown at the foot of the table. A measure of inequality of distribution of GGP among the regions (the variance) shows South Africa nearer to the upper extremes on an international scale than to the levels of European countries (Roux, 1992, p. 213). The index fell between 1970 and 1980 but then rose slightly.

Table 3.5. Regional Distribution of Gross Geographical Product, 1970–87

Region	Percentage distribution 1987 Population	GGP	Per capita GGP index 1970	1980	1987
A	9.14	12.63	1.56	1.32	1.38
B	2.88	2.83	0.78	1.14	0.98
C	6.88	5.80	0.88	1.06	0.84
D	12.76	7.20	0.57	0.48	0.56
E	24.63	14.30	0.60	0.54	0.58
F	5.83	8.35	0.63	1.33	1.43
G	11.97	3.77	0.30	0.33	0.32
H	21.09	39.90	2.10	1.81	1.89
J	4.82	5.23	0.87	1.46	1.09
Variance			0.61	0.55	0.56
Metropolitan	39.28	61.94	1.80	1.55	1.58
Homelands	35.72	4.72	0.10	0.10	0.13
Rest	25.00	33.35	0.84	1.26	1.33

Source: Roux (1992, pp. 212–3). Note that the GGP per capita index is percentage deviations above (greater than 1.00) and below (less than 1.00) the national average; the variance is the weighted coefficient of variation.

A strategy of regional equalization is important in itself and also as a means towards racial equalization. This involves not only achieving a better balance between the distribution of population and productive resources (via decentralization of production or concentration of population) but also the equalization of public service expenditure or standards. As with racial inequalities, the existing regional inequalities impose severe constraints on equalization. Even without reincorporation of the 'independent' homelands, there will be vast regional disparities to be narrowed. The ANC draft Bill of Rights makes reference to the diversion of resources from richer to poorer areas to achieve a common floor of rights for the country as a whole. Hence the importance of regional delimitation and fiscal relations in constitutional negotiations (Maasdorp, 1993). There are obvious implications for social justice in the conflict between the ANC's desire for a centralized redistributive state and that of other parties seeking greater dispersal of power in a federal structure capable of protecting regional privileges in living standards.

The same issue arises at the intra-metropolitan scale. The redistributive possibilities of local government restructuring were revealed to a limited degree before the collapse of apartheid in the Regional Services Councils incorporating both poor African townships and affluent white areas (Lemon, 1992). However, smaller local government jurisdictions with a high degree of autonomy could protect local white privilege, thereby frustrating the sub-national scale of redistribution, via local fiscal transfers, on which the townships and shacklands depend for improved service provision.

A major consideration for social justice in the city is the extent to which the built form and the peculiar land-use practices of the apartheid era constrain change towards more egalitarian structures. It is not simply that apartheid created highly segregated cities in which the patterns established by racial Group Areas will take many years to erase. It is also the case that the large income differential between the races generated a highly differentiated housing stock which will tend to consolidate differences among people manifest in class identity, even if the significance of race becomes blurred. The spatial form of the apartheid city, like other structural inequalities, will take a long time for a new society to transcend (for further discussion, see Lemon, 1991; Smith, 1992a; Swilling et al., 1991).

THE PROSPECTS FOR SOCIAL JUSTICE

The scenario for change which prevails in business circles eschews radical social transformation, instead favouring redistribution accomplished under market forces by means of accelerated export-led economic growth. More black workers would be drawn into an expanding labour force, their enhanced living standards being achieved (implicitly) without eroding those of the whites: a version of the familiar development strategy of 'redistribution through growth'. A problem with this scenario is that, even if it can be achieved, faster economic growth by no means guarantees accelerated redistribution. The policy discourse in establishment circles links public spending with inflation and stresses the 'new right' panaceas of market mechanisms, privatization and deregulation rather than state responsibility for the poor. The economic power at the disposal of large private vested interests enables them greatly to influence debates by funding and publishing research supportive of market-led reforms and opposed to a more proactive state.

In the face of such formidable propaganda, added to the former minority government's control of national broadcasting media, the voice of the former liberation movement seemed muted. The ANC now embraces a 'mixed economy', in which national development planning will avoid 'commandist or bureaucratic methods' (African National Congress, 1992, p. 11). While recognizing that market relations are an essential component of a mixed economy, the ANC 'does not believe that market forces alone will result in anything but the perpetuation of existing disparities in income and wealth' (p. 11). The actual balance envisaged between private enterprise and state control is far from clear.

What is evident is that, if post-apartheid South Africa is to make sustainable claims to social justice, redistribution has to take place on a larger scale, at a faster pace and under greater structural change than is currently envisaged in white political circles and in sections of the academic community. There also has to be much greater attention to social justice at a theoretical level, to augment the prevailing pragmatism. Perspectives such as John Rawls's liberal egalitarianism, introduced earlier in this chapter, are providing some scholars with a framework within which to couch debates on redistribution. For example, Hofmeyr and McLennon (1992, following Simkins, 1986) consider its application to education. Rawls's principle of equal access and entitlement to social primary goods entails equality and impartiality in public spending on education. Furthermore: 'extra compensatory monies should be invested in the education of historically deprived groups and affirmative action programmes should be allowed. The education system, like every other system, would therefore be so designed as to maximise the prospects of the least well-off members of society' (Hofmeyr and McLennon, 1992, p. 181). They conclude that 'South Africans will need to develop Rawls' mutual respect for a shared moral discourse' (p. 188).

There is also a need to bear in mind the breadth of the concept of social justice itself, as articulated in Brian Barry's words quoted at the outset of this chapter. The emphasis in much of the debate is on distribution, on who gets what, and where, and how this might be changed for the better. This approach is encouraged by the conspicuous inequality in people's material living conditions. The danger is, of course, that of focusing on pattern to the neglect of process, with process itself construed in the limited terms of overcoming practical obstacles to redistribution.

Preoccupation with distribution characterizes much, if by no means all, discussion of social justice in moral philosophy as well as in geography. The implications are explained by Young (1990, p. 25) as follows:

> The distributive paradigm implicitly assumes that social judgements are about what individual persons have, how much they have, and how that amount compares with what other persons have. This focus on possession tends to preclude thinking about what people are doing, according to what institutional rules, how their doings and havings are structured by institutionalized relations that constitute their positions, and how the combined effect of their doings has recursive effects on their lives . . .
>
> Rights are not fruitfully conceived as possessions. Rights are relationships, not things; they are institutionally defined rules specifying what people can do in relation to one another. Rights refer to doing more than having, to social relationships that enable or constrain action.

To Young, social injustice concerns the domination and oppression of one group or groups in society by another, not merely the distributional outcomes. This involves not only formal institutions, but also the day-to-day practices whereby subordination and exclusion are realized and reproduced as what is sometimes termed 'cultural imperialism'. Democracy is central to social justice, but formal political rights conceived merely as 'one person one vote' are not sufficient for the full democratization of society. With her mind on Rawls as well as on social reality, Young (1990, p. 116) asserts: 'Instead of a fictional contract, we require real participatory structures in which actual people, with their geographical, ethnic, gender, and occupational differences, assert their perspectives on social issues within institutions that encourage the representation of their distinct voices.' Nothing could be more pertinent to the singularly undemocratic, 'pluralistic' society of South Africa, where the inclination of the 'pigmentocracy' has been to separate and silence others.

The problem is that, if we consider the kind of principles of social justice proposed by Rawls, the obstacles which have to be faced go far beyond the practicalities of redistribution. And they are more complex than the questions of measurement implicit in defining people's basic needs or subsistence rights, for example. What is required also goes well beyond the formal democratization of such institutions as big business and the mass media—important though this certainly is if hitherto silent voices are to be heard, listened to and influence what is done. And it goes beyond reparation—the positive discrimination and affirmative action required to compensate for past injustice and give substance to the principle of equal opportunity—important though the latter is symbolically as well as practically. What is needed is a cultural revolution: a fundamental change in the way people behave towards one another, signalled especially by their (racial) appearance, in order to dissolve the dichotomy of (white) selves and 'the others' through which a profound sense of difference if not superiority has for so long been sustained.

But, this does not mean that theorizing about distributive justice in South

Africa is a mere intellectual indulgence. There is an important role to play, helping to highlight key problems in achieving change in the direction of social arrangements that could be defended from a moral point of view.

The central problem (paradox or contradiction) facing South Africa, when theory confronts reality, may be framed in different ways. One way is to stress that (as hinted earlier) the starting conditions must be right, in a moral sense, if they are to generate just outcomes; yet to turn apartheid's highly unequal distribution of income, wealth and other resources relevant to human life chances into something resembling a just starting condition will take many years. There is no clean slate, no pristine state of nature, on which to construct a new society, only the sorry reality of the here and now. The built form of the highly unequal cities is a vivid case in point. Yet as long as some of the apartheid distribution of income and wealth survives, with its unequal effective demand driving the markets for goods and services, injustice will be perpetuated, as it will be in the surviving features of the apartheid city.

Faced with these difficulties, we might be tempted to retreat from the egalitarianism implied by a 'non-racial' society. We might even argue that the perpetuation of white privilege could be in the interest of blacks. Rawls would allow us to approve such inequality if demonstrably in the interests of the worst-off. While such an argument could easily degenerate into defence of the status quo, it is by no means implausible that the economic growth on which redistribution depends requires incentives for whites to stay and work enthusiastically for the new South Africa. The responsibility of today's population for future generations is a difficult moral issue, but it is not unusual to propose present sacrifice for future gains: in this case the (temporary) sacrifice of full racial equality for enhanced redistributional possibilities. The potential for dispossessed and alienated whites to destabilize the economy and society, to everyone's disadvantage, could be part of such a calculation.

Another way of expressing the central problem is in terms of structures rather than (re)distribution. Even the most optimistic scenarios underline the limits to the reforming (no longer revolutionary) capacity of a post-apartheid government, within a realistic set of economic and political constraints. If a socialist society is no longer on the agenda, then the major private economic interests which supported and profited from apartheid will still effectively control much of the new South Africa. These include some agents responsible hitherto for legitimating racial domination, within a culture festering on unequal difference and exclusion of others rather than seeking cohesion while respecting difference and celebrating heterogeneity. If structures of domination and oppression continue to marginalize substantial, often racialized, groups of people in much richer and supposedly democratic capitalist countries, such as Britain and the United States, why should the new South Africa be any more inclusive?

The most likely scenario is profoundly depressing. It is that South Africa will steadily come more closely to resemble a normal capitalist society, its inherited racial inequalities interpenetrated by class. Its main distinguishing feature will be the size of its 'underclass', detached from the consumption norms of the well-to-do and increasingly alienated. While the rural poor may be too scattered and insufficiently organized to pose much of a threat to social order, urban elements within the underclass may become violently rebellious and vigorously repressed. In so far as this kind of situation will not be unfamiliar to the governments of the United Kingdom and the United States, not to mention other less scrupulous regimes, South Africa's speedy welcome back to the 'community of nations' is assured, as already demonstrated on the sports fields and by its return to the Commonwealth. If anything more than the formal abolition of apartheid along with universal franchise is expected, it is likely to fall well short of truly non-racial citizenship with real equality of opportunity for all. Indeed, for the mass of poor blacks, the long coveted vote may not amount to much in terms of effective power to improve their material conditions. Unless this kind of scenario proves to be unfounded, to defend such a society on grounds of social justice may require almost as much ingenuity as under apartheid.

ACKNOWLEDGEMENTS

Parts of this chapter are a revised version of an article originally published in *Area* (Smith, 1992b); it also draws on material in Chapter 8 of *Geography and Social Justice* (Smith, 1994). The author is grateful to the Departments of Geography at the University of Durban-Westville and the University of Natal, Durban for supporting a visit to South Africa in 1991 to undertake research which contributed to this chapter, to the Hayter Fund of the University of London for a grant towards travel expenses and to the London office of the African National Congress for their encouragement.

REFERENCES

African National Congress (1992) *Discussion Document: Economic Policy*, Department of Economic Policy, ANC, Johannesburg.
Barry, B. (1989) *Theories of Justice*, Harvester-Wheatsheaf, London.
Bond, P. (1991) *Commanding Heights and Community Control: New Economics for a New South Africa*, Raven Press, Johannesburg.
Claassens, A. (1991) *Who Owns South Africa: Can the Repeal of the Land Acts De-racialise Land Ownership in South Africa?* Occasional Paper 11, Centre for Applied Legal Studies, University of the Witwatersrand, Johannesburg.
Davis, D. (1992) 'Taxation in post-apartheid South Africa' in R. Schrire (ed.), 1992, 105–20.
Eckert, J. (1991) 'National dialogue: towards an ethics charter', *Indicator South Africa*, 8(2), 44–8.

Hofmeyr, J. and McLennon, A. (1992) 'The challenge of equalizing education' in R. Schrire (ed.), 1992, 174–92.

Kymlicka, W. (1990) *Contemporary Political Philosophy: an Introduction*, Clarendon Press, Oxford.

Lemon, A. (ed.) (1991) *Homes Apart: South Africa's Segregated Cities*, Paul Chapman, London; David Philip, Cape Town and Indiana University Press, Bloomington.

Lemon, A. (1992) 'Restructuring the local state in South Africa: Regional Services Councils, redistribution and legitimacy' in D. Drakakis-Smith (ed.), *Urban and Regional Change in South Africa*, Routledge, London, 1–32.

Maasdorp, G. (1993) 'Drawing the lines: economic implications of regional restructuring in South Africa', *Optima*, 23(9), 25–9.

Marcus, T. (1991) 'Palace coup on land reform', *Indicator South Africa*, 8(4), 49–54.

McGrath, M. and Holden, M. (1992) 'Economic outlook 1981–1992', *Indicator South Africa*, 9(4), 35–40.

Miller, R.W. (1992) *Moral Differences: Truth, Justice and Conscience in a World of Conflict*, Princeton University Press, Princeton.

Moll, P.G. (1990) *The Great Economic Debate*, Skotaville Publishers, Braamfontein.

Moll, P.G. (1991) 'Conclusion: what redistributes and what doesn't' in P. Moll, N. Natrass and L. Loots (eds), (1991), 118–34.

Moll, P., Natrass, N. and Loots, L. (eds) (1991) *Redistribution: How Can it Work in South Africa?*, David Philip, Cape Town.

Moore, B. (1992) 'The case for a land tax: from entitlement to restitution', *Indicator South Africa*, 9(2), 25–9.

Nozick, R. (1974) *Anarchy, State, and Utopia*, Basic Books, New York.

Peffer, R.G. (1990) *Marxism, Morality, and Social Justice*, Princeton University Press, Princeton.

Rawls, J. (1971) *A Theory of Justice*, Harvard University Press, Cambridge, Mass.

Reekie, W.D. (1990) 'Privatisation and the distribution of income' in F. Vorhies (ed.), *Privatisation and Economic Justice*, Juta & Co., Cape Town, 4–15.

Roux, A. (1992) 'Regional options for equity and efficiency' in G. Howe and P. le Roux (eds), *Transforming the Economy: Policy Options for South Africa*, Indicator South Africa, University of Natal and Institute for Social Development, University of the Western Cape, 211–28.

Sachs, A. (1990) *Rights to Land: a Fresh Look at the Property Question*, Institute of Commonwealth Studies, University of London.

SAIRR (1989) *Race Relations Survey 1988/89*, South African Institute of Race Relations, Johannesburg.

SAIRR (1992) *Race Relations Survey 1991/92*, South African Institute of Race Relations, Johannesburg.

SAIRR (1993) *Race Relations Survey 1992/93*, South African Institute of Race Relations, Johannesburg.

Schrire, R. (ed.) (1992) *Wealth or Poverty: Critical Choices for South Africa*, Oxford University Press, Cape Town.

Simkins, C. (1986) *Restructuring South African Liberalism*, South African Institute of Race Relations, Johannesburg.

Smith, D.M. (1990) *Apartheid in South Africa*, third edition, UpDate series, Cambridge University Press, Cambridge.

Smith, D.M. (ed.) (1992a) *The Apartheid City and Beyond: Urbanisation and Social Change in South Africa*, Routledge, London.

Smith, D.M. (1992b) 'Redistribution after apartheid: who gets what where in the new South Africa', *Area*, 24 (4), 350–58.

Smith, D.M. (1994) *Geography and Social Justice*, Basil Blackwell, Oxford.

South Africa (1991) *White Paper on Land Reform*, Government Printing Office, Pretoria.

Swilling, M., Humphries, R. and Shubane, K. (eds) (1991) *Apartheid City in Transition*, Oxford University Press, Cape Town.

van der Berg, S. (1991) 'Redirecting government expenditure' in P. Moll, N. Natrass and L. Loots (eds) (1991), 74–85.

van der Berg, S. (1992) 'Social reform and the reallocation of social expenditure' in R. Schrire (ed.) (1992), 121–42.

Waldron, J. (1992) 'Superseding historic injustice', *Ethics*, **103** (1), 4–28.

Young, I.M. (1990) *Justice and the Politics of Difference*, Princeton University Press, Princeton.

4 Land Relations and Social Dynamics: Reflections on Contemporary Land Issues in South Africa, with Particular Reference to the Eastern Cape

TONY BUCKLE

For the great majority of South Africans, the effective resolution of the land question is a key element of any post-apartheid dispensation. Apartheid, and the colonial and segregationist policies that preceded it, acted to displace the black population from nearly 90 per cent of the land area of South Africa and to place this land in the hands of the white elite. To understand the full extent of the impact that this mass displacement had on the black population, it is necessary to explore both the structure of land tenure in South Africa and the complex values attaching to land in this country. This chapter focuses on these dimensions of the land question and assesses their impact on the relationship between land, property and power. Starting with a brief review of some of the theoretical considerations concerning property and land tenure, it proceeds to an investigation of the impact of land relations upon the social dynamics of two communities in the Eastern Cape. It concludes by examining some themes which require careful consideration if the post-apartheid government is to address the land question effectively.

LAND, PROPERTY AND POWER: A THEORETICAL INTRODUCTION

Property is a series of rights and duties that posits the individual in a specific relation *vis-à-vis* others with respect to resources within a particular domain. It exists at a number of levels, both within a social group (family, community, society) and without it and acts as recognition that the property interest of one party is protected by a right only when others fulfil their duty to respect that right (Bromley and Cernea, 1989). Seen in these terms, it is apparent that 'property . . . is not a relation between people and things . . . [but] a relation between people,

The Geography of Change in South Africa, edited by Anthony Lemon.

concerning things' (Ferguson, 1990, p. 142). As such, the property relation is essentially a power structure, empowering rights holders by granting them exclusive access to owned resources while disempowering non-rights holders by denying them access to owned resources. At the society level, property relations empower some sectors of society at the expense of others.

Under capitalism, the social structure derives primarily from the mode of production, and accordingly, the property system reflects the dynamics of the capitalist mode of production in the first instance. Production and the creation of economic surplus dictate the values and functions attaching to the object world, and therefore objects become commodities, evaluated primarily in terms of their productive potential.

Similarly, property relations are assessed in terms of their capacity to secure and enhance the productive potential of commodities, since the purpose of property under capitalism is to enable the individual to produce. The principal property regime under the capitalist mode of production is private property, an institutionalized system of rights and duties, founded upon the authority of the state and legislature. Private property rights include the rights of purchase, sale and transfer and consequently accommodate both inequalities in the level of property ownership between people and their complete alienation from the object world. These are distinct advantages to the capitalist economy because they allow production for accumulation.

It is also argued that private property ownership stabilizes the capitalist system by providing property owners with the security of a stake in the system. This is held to be particularly true of private land ownership: the socio-spatial stake provides owners with both tangible proof of their place in society and a sense of belonging to society. Thus, Bellman (1927) argues that 'the man who has something to protect and improve—a stake of some sort in the country— naturally turns his thoughts in the direction of sane, ordered and perforce economical government. The thrifty man is seldom or never an extremist agitator. To him revolution is anathema' (p. 54).

Similar arguments have been put forward by some commentators in contemporary South Africa: 'If we could give them [the deprived] a genuine stake in the community and pride in owning their own home, it is unlikely they would throw it up by responding to the call from the township warlords' (Swalens, 1990).

The relationship between land, property and power under capitalism, therefore, exists at two distinct levels. First, private property relations in general enable individuals to secure for themselves any surplus yielded by the land as a commodity. Second, the private ownership of land acts to stabilize the capitalist system and ensure its reproduction. At both levels, property and power are integrally connected; property relations are seen as the foundation upon which the productive process, the essence of the capitalist system, is constructed.

It is evident that under capitalism, the structures of property are founded upon a particular form of societal organization. Where the configuration of social forces is different, it would be expected not only that the structures of property differ but that the very meaning of property varies. Different cultures project different 'ways of seeing', evaluating the same resource differently. In advanced capitalist countries, the evaluation of land reflects the emphases of the prevailing mode of production and is primarily economic. However, in Africa, the appraisal of land reflects markedly different values.

According to Cohen (1970), the prevailing common feature of traditional African societies is 'the emphasis on social relations as the primary value' (p. 39). Traditional African value systems focus upon the collective, with constituents deriving their individual identities from their relations with the greater society. For Sparks (1990), such societies exhibit a 'participatory humanism', wherein the value of individuals stems from their participation in, and contribution to, broader society, an ethos captured in the Xhosa proverb '*ubuntu ungamntu ngabanye abantu*'—people are people through other people.

An appreciation of such values is crucial to an understanding of property systems in South Africa, not only historically but also in the present. Property structures derive from social relations rather than production, reflecting and informing social rather than economic values in the first instance. The consistent emphasis within production for example, is upon the fulfilment of subsistence need rather than the accumulation of surplus.

This social ethnic permeates all values attaching to land, with the result that most of the principles involved in land rights relate to social relationships articulated through land holding rather than to economic relationships concerning production. The focus is social security. Thus, the granting of land acts to create ties between the families involved with the result that households inhabiting what was once one land holding tend to form a co-operating territorial group, maintaining internal solidarity and fulfilling subsistence needs. The land expresses group membership and forges the social relations vital to survival strategies.

Similarly, the land ethic affords landholders a considerable degree of tenure security. It rules that allocated plots cannot be taken away without due process and compelling reason because land is the foundation of a family's subsistence existence. Accordingly, the property system cannot confer the rights of alienation and transfer because individuals can no more alienate their need for land than they can their need for food. With land so essential to life, the social land ethic dictates that all individuals must have inalienable access to it.

Cross (1991b) draws out three main features of the African social land ethic. First, and most fundamentally, all indigenous tenure systems recognize the right of all resident families to make a claim on the community for land. Forms of tenure that provide land only to some people and not to others are seen as essentially inequitable and unjust. Second, a land holding is assumed to include every resource needed to make the family self-sufficient, encompassing a residential site and land for both cultivation and grazing. Free access to a whole package of environmental resources (water, firewood, thatching grass, clay etc.) is also assumed; historically, these have been communally owned despite the geographical specificity of their occurrence. Third, landless families have a claim on the community to provide them with land. A family with more land in excess of its perceived requirements can be asked to provide for those in need, and although landholders can refuse such requests, refusals can only be justified in terms of other social needs.

In Africa, land not only sustains society in a biological sense, it also plays an important support function outside the spheres of production and distribution. The universal land right ensures that all individuals have a tangible stake in society, and this forges a binding collective consciousness acting to stabilize society, instil a sense of belonging and ensure social reproduction. Furthermore, the social land ethic affords individuals status for producing for others, accommodating the establishment and fulfilment of client relations and obligations.

Cross (1991b) also sees the social land ethic as a mechanism to balance providing land to the landless against upholding possession by land holding families. Ranking in land rights according to the chronology of settlement is recognized but at the same time as the community's obligation to the poor. The land right trades off universality against hierarchy: it ensures that there is both a place for everybody but also that everybody is in his/her place. As a result, land holding both reflects and underpins the social structure of communities with the result that it is considered natural that those descent groups longest settled in an area will have the largest land holdings and will also provide leadership and authority. In a very direct way therefore, land holding expresses power.

In summary, it is clear that property systems reflect the values and demands of the societies in which they are constituted. Under capitalism, the private property ethic reflects the dominance of economic values over social values, while in Africa, the traditional land ethic reflects the dominance of social values over economic values. Under both systems, however, the relation between land, property and power is central to societal constitution and dynamics.

BACKGROUND TO FIELDWORK

What follows is a brief overview of the way in which land issues have informed the past and present development of Mgwali and Lesseyton, two former mission communities in the Border Corridor, Eastern Cape Province (Figure 4.1). The mission station is a revealing arena for a study of land issues in contemporary South Africa due to its merging of the two markedly different property regimes and cultures described above.

In both communities there are several distinct land relations. First, there are the quitrent title holders, typically direct descendants of early mission converts who were awarded personal legal title to a residential site and an arable field as

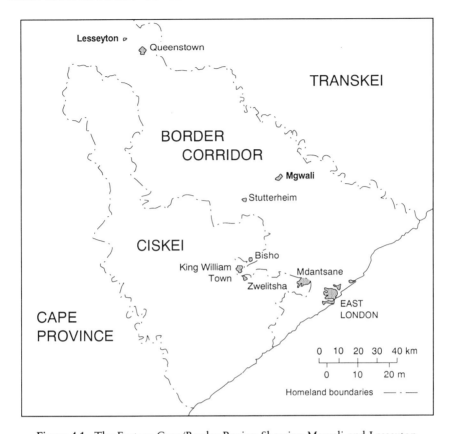

Figure 4.1. The Eastern Cape/Border Region Showing Mgwali and Lesseyton

well as access to the communal grazing land. Second, there are holders of certificates of occupation, whose land documents were granted much more recently and which extend only to a residential site. Third, and by far the most numerous, there are the landless, typically evictees from white farms or refugees from Transkei or Ciskei.

LAND, PROPERTY AND POWER IN MGWALI

Established in 1857, Mgwali evolved to become one of the most progressive and prominent mission stations in the Eastern Cape. It was the first mission station in the whole of Africa to have an indigenous black minister, Tiyo Soga, and as the parish of the Rharabe royal family, it soon achieved a significance disproportionate to its numerical size.

Land relations are integral both to the historical development and present dynamics of Mgwali. As Slater (1980) notes, the active manipulation of socio-space was central to colonial Native policy. Both church and state cast individual land ownership as the key to community passivity and loyalty. Successive Mgwali ministers cast private land ownership as the foundation of market production and heightened individual productivity, a view which the mission residents rapidly adopted themselves. This pressure to extend private property relations to worthy congregants eventually prompted the state to grant 152 quitrent titles to Mgwali residents in 1873.

However, in the Cape Colony, the advantages of private property were not confined to socio-economics. Under the Colony's property-based franchise, only individuals owning land could vote in Colony elections. Furthermore, under the 1881 Villages Management Act, only registered voters could elect, and be elected on to, the Village Management Boards instituted in non-Reserve black communities such as mission stations. This situation persisted until the abolition of black voting rights in 1936 with the result that for over half a century only Mgwali's title-holding residents had a political voice.

The Mgwali title holders evolved to become prominent members of the African peasantry (see Bundy, 1979). Tragically, their very progress made them prime targets for apartheid's policies of racial limitation. Apartheid was in many ways a reversal of colonial land policy, intended to 'return' black society to communal land practices and authority structures through the active manipulation of space.

The apartheid plan for Mgwali was to remove the community *in toto* to Ciskei. However, the Mgwali community resisted removal with a determination that had few parallels in South Africa. The resistance campaign was orchestrated by the Mgwali Residents Association (MRA), whose leadership throughout the resistance campaign comprised almost exclusively prominent Mgwali landowners, individuals whose status and power was integrally wrapped up with the specific socio-spatial constitution of Mgwali. The private property ethic can be clearly discerned as a primary motivating force for the landowners' resistance. First, the landowners had strong economic reasons for wanting to remain in Mgwali. As quitrent holders, they each commanded personal ownership of a sizeable residential and garden plot, an arable field and the right to run their cattle on the common pasturage. This not only secured them a smallholder income in the present but had in the past afforded a solid financial foundation, which over generations had enabled many landowners to diversify their entrepreneurial activities and establish small businesses such as shops and taxis.

Second, the landowners had strong social reasons for resisting the removal. They lived well in Mgwali, enjoying large brick houses, amenities including schools, a clinic and a church and an unusually high degree of community stability. Furthermore, they enjoyed an independence from tribal structures due to their independent legal ownership of land. The title holders were the embodiment of the colonial project to release individuals from tribal sanction via independent land ownership. However, the title holders were well aware that this independence was founded on the specific tenure circumstances of Mgwali and would not be maintained in Ciskei, where, as Cokwana notes (1988), '[quitrent] title may be forfeited under any of a number of conditions, such as failure to beneficially occupy the allotment, non-payment of quitrent, or conviction of certain crimes' (p. 310). Once removed to Ciskei, the Mgwali landowners would have to pay due deference to the state for fear of forfeiting their lands.

Third, the landowners stood to lose their political clout if they were removed. As the local landowners, they assumed a natural stature and importance in the community which was founded upon their historical control of local production. This assumption of authority is evident from the fact that the MRA achieved legitimacy in Mgwali despite being self-appointed.

These material considerations were reinforced by the values stemming from the African social land ethic. The connection between space and society is starkly evident in Mgwali, where the community's socio-spatial order expresses landholding differentials, chronology of settlement, kinship affiliations and so on. This constitutes a further dimension to power relations in Mgwali, in which everyone has their (social and spatial) place, and everyone knows their position vis-à-vis others. The socio-spatial order is a concretion and legitimation of the power structure in Mgwali, and the landowners were aware that removal would undermine their position in the community.

Furthermore, the land ethic emphasises the link between the contemporary community and past generations with the result that attachment to place is very deep and personal. As Tuan (1974) notes, this connection to a specific place 'is of a religious nature. The tie is one of kinship, reaching back in time from proximate ancestors to distant semi-divine heroes' (p. 242). During the removals era, the MRA highlighted not only Mgwali's mission past but also the community's ties to the Rharabe warrior-king Sandile. To relinquish Mgwali would be to sever connections with this past and to forfeit the community's heritage. The land ethic ensured that the resistance struggle focused upon much more than mere material interests.

Significantly, the distinct land orientation of the resistance campaign in Mgwali presented its own set of problems. Mgwali's squatters were acutely aware of their landless circumstance, and accordingly, their priority was to obtain their own land, either within Mgwali or outside it. As the landowners and the MRA were determined to maintain the integrity of land holding in Mgwali, the squatters were obliged to look elsewhere for sites, leaving them receptive to promises of land in Ciskei. As the squatters formed a numerical majority, the title holders feared that Ciskei would manipulate the squatters into voting for the community's resettlement in the homeland. As a result, the MRA made frequent attempts to incorporate the squatters into its structures, providing them with food and employment in community projects and involving their children in their youth sports programme. These incorporationist strategies aimed to foster unity in resistance to removal, but it is significant that throughout the removals struggle the MRA at no stage attempted to win over squatter support via site allocation. The title holders were fighting to preserve their position within the community, and changing the socio-spatial order to accommodate the squatters would have compromised this hierarchy as surely as removal.

In the present, land relations remain central to societal dynamics in Mgwali. Land relations remain actively determinant of societal differentiation, particularly in terms of symbolism and attitudes. Many of the title holders still assume, and are still assumed to have, wealth, power and authority in the community on account of their quitrent titles. That this is the case is evident from squatters' explanation of the title holders' perceived material advantage, which is attributed to the title holders' ownership of arable fields. Consequently, arable fields are still cast as generators of wealth, despite the fact that there has been virtually no ploughing in recent years due to a combination of drought and inadequate investment, and although the landowners and squatters alike are materially dependent upon pensions and remittances from migrant workers.

Land relations also continue to dominate a community consciousness which posits the title holders as the ultimate community reference group and in many ways reflects their interests. There is a distinct equation of land ownership and empowerment, and to own land, particularly to have quitrent title, is regarded as the key to power, privilege and prestige. The relation exists at a number of levels. First, land is perceived to afford to the owner a secure economic foundation, not only because of the land's intrinsic productive potential but also because formal land ownership brings with it an official livestock card and the legal right to run livestock on the commonage. Second, title provides a secure social foundation, protecting landowners from removal either by the state or by other landowners and, accordingly, is the foundation of considerable personal independence from the will of others. Furthermore, legal ownership papers seem to symbolize recognition by the state and the community that the individual is a true resident of Mgwali.

The land–power relation is expressed in various practices and attitudes in the contemporary community. Most stark is the title holders' ability to manipulate the settlement to their advantage. Landowners settle squatters on unoccupied sites to reserve them for their children, evicting the squatters once their children come of age. Consequently, small-scale removals are frequent in Mgwali with one squatter complaining of being ejected nine times from different sites.

More often the superiority of the title holders is simply assumed. Non-title holders felt that 'we are treated differently because we don't plough', noting how they were called 'voortrekkers' by some of the landowners and did not think of themselves as real residents of Mgwali because of their landlessness. The squatters refer to the landowners as 'bosses', explaining that they often require the landlords' permission to do things and regarding themselves as being 'under the control' of the individual or family who owns the site on which they reside.

The title holders, for their part, are extremely patronizing towards the squatters whom, in their view, they have allowed to come and settle in Mgwali. It is significant that all residents frequently refer to themselves and others in terms of a land relation—'quitrents', 'certificates', 'squatters'—demonstrating how land relations are at the root of community division and consciousness.

These divisions are fundamental to the socio-spatial hierarchy in Mgwali. This comprises multiple gradations: quitrent title holders; certificate of occupation holders; farm evictees who have been allocated their own sites but not certificates; farm evictees who have been settled on sites 'under' landowners; farm evictees living with other families on single sites. The land ladder is complicated by kinship ties, land grants derived from church membership, history of settlement and other factors, but despite its complexity, most villagers are extremely aware of their own position within it, and this is reflected in the immediate priorities of different land groups. The certificate holders' priority is quitrent title: 'our main concern is for the evictees to own arable land together with us so that we can all be like the quitrent'. Sited evictees have a different priority: 'we don't have certificates of occupation . . . our problem is very much our certificates of occupation'. For the siteless, the priority was again different: 'the first thing that we want is a place to stay'. This group's problem is that their families 'stay with other families', and accordingly, their primary demand is that their families have 'their own place to stay'. However, although the priorities vary, in the longer term all residents aspire to be quitrent title holders. To achieve status in Mgwali, an individual needs to own land.

The land ladder is a distinct power structure, positing the quitrent title holders on top and extending into a chain of dependency relations, articulated through land, which underpin community dynamics. For not only are many of the landless villagers acutely dependent upon landowners for residential and garden sites but also for all other land rights: the right to grazing, the right to cultivation, even the right to residence. The whole farming effort in Mgwali depends upon the will of the title holders, and many non-title holders express dependence on the title holders for a host of other activities, from the construction of homes through to the veneration of ancestors. This power structure seems to be accepted as quite legitimate by the community. It appears that land acts as a form of symbolic capital (Harvey, 1985), its skewed distribution legitimizing the skewed distribution of political power in Mgwali, making the established order appear natural.

As most of the community's demands for development and empowerment focus upon land, it is unsurprising that land issues are the primary focus of local politics in Mgwali. However both the MRA and the local ANC appear powerless to compromise the landowners' control of land matters in the community with the result that the title holders' position of power in Mgwali is only confirmed and reinforced. With community politics unable to address the primary issue of land ownership, the majority of Mgwali residents are becoming increasingly frustrated and disillusioned.

LAND, PROPERTY AND POWER IN LESSEYTON

Established in 1848, the Lesseyton mission station quickly developed into a stable and Westernized black community and achieved prominence during the 1851 Frontier War as an outpost of 'loyalism' (see Bundy, 1979) in a rebellious region. The mission comprised a fertile valley and lent itself to both arable cultivation and sheep herding and the adoption of Western-style farming techniques.

As in Mgwali, land relations were central to the way in which the Lesseyton mission developed. The mission lands were formally granted to the Methodist Church in 1874, and the Church arranged the extension of quitrent title to its 32 most prominent congregants in 1876. This secured the distinct political and economic advantages outlined above to the chosen title holders, prompting the development of distinct material divisions within the community based on land.

The extension of quitrent title also established the distinction between the 'Village' and the 'Trust' in Lesseyton. The quitrent plots were demarcated in a grid pattern near the mission church, and the settlement there was re-ordered into a formal Village. Settlement in the Trust outside the Village remained informal, structured along traditional rather than colonial lines. This spatial distinction informed social consciousness, and the Village title holders came to view themselves as distinct and superior to the residents of the Trust. This view was reinforced by the mission church's decision to grant the Lesseyton Villagers dominion over the Trust.

Like the other Border 'black spots', Lesseyton was considered 'badly situated' by the architects of apartheid, and its inhabitants scheduled for removal to Ciskei. Removal was fiercely resisted by all sections of the Lesseyton community for reasons grounded both in absolute space and relative space. In terms of absolute space, Lesseyton was an extensive and fertile valley, supportive of both agriculture and pastoralism, and far superior farmland to the proposed resettlement area in Ciskei's arid Hewu district. In terms of relative space, Lesseyton was near Queenstown, the major town of the region. Queenstown not only provided a market for farm produce and extensive shopping facilities but was also a major labour market for which Lesseyton was a convenient dormitory settlement.

However, both the squatter and landowner groups also had their own particular reasons for resisting removal, deriving from the peculiarities of land relations in Lesseyton. In 1948, the Methodist Church sold the Trust area of the mission station to the South African Native Trust. This sale undermined the title holders' authority over the Trust, effectively liberating the Trust community from their Village overlords and allowing them to take control of their own affairs. This meant that during the removal struggle, both the Lesseyton title holders and the Lesseyton squatters were fighting to defend tangible socio-spatial 'stakes' in the community, even though these 'stakes' were markedly different, indeed mutually exclusive. The title holders were struggling to protect their claim to the whole of Lesseyton, both Village and Trust. The squatters, in contrast, were struggling to protect their claim to the Trust. However, despite their incompatibility, both struggles were founded upon a common land ethos, which focused on the strong, positive relation between land and personal power in the community hierarchy. To control land, particularly to own it, meant power; conversely, landlessness equated to powerlessness.

The particular stake in the community that the title holders were seeking to defend derived from their titles. As in Mgwali, the title holders assumed power over the Lesseyton community on account of their legal ownership of the land. Realizing that removal to Ciskei would jeopardize this 'stake', they were keen to defend their source of power and privilege. The squatters, on the other hand, were seeking to defend their independence from landowner control, achieved as a result of the state's purchase of the Trust. Furthermore, with Lesseyton situated within the Republic of South Africa, both squatters and title holders were seeking to defend their independence from bantustan tribal authorities.

The Trust itself conferred distinct advantages on the squatter population. It is a sizeable and productive tract of land, capable of supporting both intensive subsistence gardening and extensive grazing. Furthermore, through its proximity to the Queenstown labour market, it afforded the squatters a degree of economic independence from the land that residents of the vast majority of other rural black communities did not enjoy. Traditionally, squatter incomes in Lesseyton have derived predominantly from urban remittances rather than from farming, giving this group economic independence from the landowners. As a result, the power of the squatters in Lesseyton stands in stark contrast to their subordination in the other Border 'black spots' and underpinned their determination to resist removal to Ciskei.

Significantly, therefore, both the landowners' and the squatters' resistance to removal from Lesseyton was essentially defensive and their orientation, although militant, essentially conservative. However, it is also evident that the community was keenly aware of the incompatibility of its two competing claims. Throughout the removals period the squatters and the title holders remained distinctly antagonistic towards one another, and despite their common interest in resisting the will of the state, the community never came together in one single resistance structure like the MRA in Mgwali.

This seems to explain why politicization remains intense in Lesseyton despite the termination of the state's removals strategy in the Border corridor in 1985. Both the landowners and the squatters are still struggling to secure their respective stakes. The landowners remain determined to re-establish their authority throughout Lesseyton, while the squatters attempt to consolidate their position within the community, particularly their control over the Trust. The common land ethos continues to underpin what is essentially a struggle for power. There remains a distinct respect for private land ownership, particularly via quitrent title, and the institutions that uphold private land ownership. Hence, despite the oppression of apartheid, the white government, the law and even the police remained remarkably legitimate throughout the Lesseyton community with both the landowners and the squatters looking to the state to intervene in the contemporary land dispute on their behalf.

This common land ethos permeates the competing claims of the landowners and the squatters. The title holders have the legal documents and consequently believe themselves to be the only legitimate residents of Lesseyton. Their most militant members have attempted to secure the removal of the entire squatter population from Lesseyton precisely on the grounds that the squatters have no legal title to the land there. They argue that as the state acts to remove squatters from privately owned 'white' farmland, it should do the same in Lesseyton. The very same land ethos underpins the squatters' claims in Lesseyton. The squatters do not challenge the landowners' claims to the Village and emphasize that the residential sites and arable fields there have never been encroached upon. They limit their claims to the Trust, basing these claims on the fact that the landowners do not have title to the Trust, as it is owned by the state. Accordingly, they are residing on public land, not private land, and are not in any way breaking the law or compromising private property rights.

With land issues at the heart of community grievance, it is unsurprising that the demands of the community residents centre on land rights. As in Mgwali, the ultimate target for all residents in quitrent title, that is, to have exactly the same land privileges as the Village landowners. The title holders still assume the position of community reference group because even though the specific local land circumstances compromise the landowners' authority over the (whole) community, title itself is perceived to confer power upon its holder.

All residents seem to perceive land control as the key to power. This is startling in the case of the squatters since in many ways it leads them to participate actively in their own marginalization. The squatters are an oppositionally defined group: the landowners refer somewhat disparagingly to all non-landowners as squatters, and it is notable that non-landowners adopt the same terminology and frames of reference despite obvious allusions to inferiority. Furthermore, the squatters largely defer to the landowners, and their feelings of powerlessness are expressed in a variety of different emotions: fear, envy, hate and insecurity. However, resentful as the squatters are about their current position in the Lesseyton hierarchy, the hierarchy and its construction are essentially accepted. It is only their relative position in the structure that the squatters want to change. This is evident in the consistent demand for title, as the major aim of the squatters is to move up the land-based social hierarchy. It is also starkly apparent in the demands of the squatters to be enabled, via the acquisition of title, to exploit other landless people in exactly the same way as the landowners currently exploit them.

Unsurprisingly, the land–power ideology has significantly impacted on contemporary community politics. The authority of any local political structure appears dependent upon its capacity to control land. This is evident from the geography of authority within the community. In the Village, all the land is owned by private landowners. Accordingly, the landowners' authority within the Village is not questioned. In the Trust, however, the land is owned by the state, not private individuals, and consequently, authority is uncertain. State-owned land does not appear subject to the same rules and respect as individually owned land, and as a result, authority within the Trust lies with those who can exercise land rights there. Hence, the authority of the squatter-based Lesseyton Residents Association (LRA) stems from its involvement, in conjunction with the state Department of Development Aid (DDA), in the formal demarcation and allocation of sites to squatters in the Trust in 1990. However, since this time, the DDA has been abolished and the LRA has been unable to authorize any further site demarcation despite community pressure on it to do so. This has prompted a multiple-level struggle for control of the Trust. The Village title holders maintain their historical argument for control of the Trust. The LRA attempts to maintain the current pattern of landholding. However, both have been unable to control landless individuals demarcating their own sites, either in and around the Trust villages, or, in the case of the iZola squatters, in one of the community's grazing camps. The latter actions not only compromise the authority of the title holders and the LRA, they also act to compromise the very integrity of land holding and use in the community. *In extremis*, this could result in the demise of 'property' in the Trust, as the rights and duties fundamental to a property regime collapse (Bromley and Cernea, 1989).

Certainly such invasions are compromising the rights associated with formal land ownership in the community. This is evident from comments made by the squatters whose sites were formalized by the DDA in 1990. This particular squatter faction now strongly and vociferously opposes any further self-demarcation in Lesseyton on the grounds that their land rights and powers are being compromised by it. This is ironic in many ways, not least since the group owes its own sites to self-demarcation. However, such politics only indicate the significance of land circumstances to political affiliations in Lesseyton.

LAND, PROPERTY AND POWER: SOME CONCLUSIONS

It has long been recognized that property is not so much a relation between people and the world of material objects as a relation between people concerning the world of material objects (Ferguson, 1990). As such, the property regime lies at the very centre of the societal fabric with each informing and being informed by the other. Accordingly where land is the primary object of property, as in much of black South Africa, land relations and power relations are effectively synonymous. Thus land relations 'form the basis of political organisation in present black-occupied areas, with the tenure system exerting a powerful influence not only over the production process but also over the constitution of society and its political order' (p. 63). This is starkly evident in homeland areas where the re-institution of 'traditional' tribal authority was founded upon the control of land. As Haines and Tapscott (1988) note of rural Transkei, 'the allocation of land is probably the most crucial mechanism for the interplay of corruption and control [as] land allocation is "traditionally" the prerogative of the chief' (p. 169). The control of land as the basis of the chiefs' power is fundamental to the reproduction of the homeland system.

Land relations are scarcely less fundamental to the societal dynamics of the two 'black spot' communities, Lesseyton and Mgwali, discussed in this chapter. The machinations of the land, property and power complex critically affected the historical development of both communities and continue to inform community structure and consciousness.

This finding should not be cast as merely a hangover from the colonial land and property manipulation. The forging of a strong and direct relationship between property and power was not a tremendously radical cultural departure for the indigenous population, especially when the property relation was articulated through land. The traditional Xhosa culture enforced a similar land–power relation, starkly evident from the expression *yinkosi umhlaba* ('the land is chief'). Although its foundation and form changed, the relationship itself was reinforced rather than invented during the colonial era and it remains informed by both the ethics of the colonial private-property regime and the traditional African social land ethic. From the private-property ethic it derives the theme that land ownership empowers the rights holder by conferring personal security, independence and legal and political recognition. From the African social land ethic comes the idea that land is vital for cultural expression and the establishment of inter-personal patron–client relations. These twin foundations explain not only the perceived power of the landowners in Mgwali and Lesseyton but also the assumed powerlessness of the landless. The landless have no means of defence against removal, no means of production from which they can secure an income and no medium through which they can express their individuality and identity.

The land–power ideology seems to have achieved hegemony in both Mgwali and Lesseyton and over time has constructed a distinctly land-based social hierarchy in each community, positing the landowners as the authority strata and community reference group and the landless as the marginalized and subordinate group. This structure is seemingly upheld by all with the effect that in a very real way the majority of residents contribute to their own marginalization. The extent to which the relation is simply assumed as natural is evident from the land demand of residents which never focused upon a change in the system of land holding but only upon a change in the relative position of the individual within the land–power structure. It was quite apparent that a change of land circumstance would merely have the effect of turning the oppressed into the oppressor.

These issues clearly have implications for the broader issues of social justice and redistribution in post-apartheid South Africa. At one level, they underline the concerns of Bond (1990) and McCarthy (1988), among others, about the impact of private-property relations in the urban sphere. For these authors, private-property ownership is being foisted upon urban blacks in an attempt to disrupt the collective consciousness engendered by shared circumstance. Private-property ownership is manipulated to achieve the acquiescence of more progressive elements to diminish the challenge of the collective. However, at a critical second level, they also raise concerns about the continuation of exploitative black-on-black power relations articulated through land. The simple extension of quitrent title to the landless for example, although empowering current residents of Mgwali and Lesseyton, would fail to address the underlying relations of exploitation which consequently may be merely reinforced.

These things said, however, it is important to note that at the level of community, the specific constitution of the land–power ideology and land-based social hierarchy are significantly influenced by local circumstance. For although the dynamics and consciousness of Mgwali and Lesseyton express many similarities, they also display large and important differences.

This is clearly the case when the relative strengths of the landowners and the landless in the two communities are compared. In Lesseyton, the squatters are in a much more powerful position *vis-à-vis* the landowners than their equivalents in Mgwali. This is due to peculiarities of land circumstances in the community. First, the South African Native Trust's purchase of the Trust in 1948 acted to break the landowners' power within Lesseyton because from that date not all of the land in the community was privately owned. The Trust residents were transformed from being squatters on the community's land to being squatters on state land and were thus freed from landowner authority. Second, there is a significant difference in the number of titles granted in Lesseyton and Mgwali. In Mgwali there are 152 quitrent title holders: in Lesseyton, there are only 32. Third, the land in Lesseyton has an important relative as well as absolute value due to its proximity to Queenstown. This gives the Lesseyton squatters a degree of economic independence from land that their equivalents in Mgwali do not enjoy. Fourth, the state's backing for the squatter-based Lesseyton Residents Association (LRA) has proved difficult for the landowners to counter. In many ways, respect for the state is an integral part of their own social consciousness, and although they attempt to circumvent this problem by distinguishing between 'Pretoria' and the Department of Development Aid, the Lesseyton landowners are obliged to accord the state-backed institution some authority.

Although local specifics have acted to compromise the position of the Lesseyton landowners, they do not seem to have undermined the prevailing land–power ideology. Indeed, in many ways they are reinforcing it. The LRA's own position of power is attributable to its perceived authority in land matters which dates back to the site demarcation of 1990. Consequently, the LRA's own power stems from the land–power ideology and the respect that is accorded to those who decide land matters. Currently, the LRA's authority is itself under challenge as a result of residents deciding to take site allocation into their own hands. Significantly, however, in all these instances the potency of the land, property and power relation remains intact. The only issue is that, currently, no one group in Lesseyton has achieved paramount authority in land matters. This contrasts greatly with the situation in Mgwali. The landowners there control all land matters and, consequently, remain firmly in control. The MRA are unable to compromise, or indeed even challenge, the authority stemming from land and are obliged constantly to consult the landowners.

The issue of land control therefore largely explains the differing politics of the two communities. In Mgwali, the landowners dominate the community politics, due to their secure control of land matters. Although there is resentment of the landowners' privileged position, particularly among the youth, they are not under political challenge. In Lesseyton, however, there exists a 'critical moment' (Crush, 1991) as no one group has achieved authority in land matters. As a result, there is intense political debate. Tragically, the absence of politics in Mgwali is depriving residents of the opportunity to vent their frustrations with the effect that a sudden outbreak of violence is more likely there than in Lesseyton.

Local specifics are crucial to an explanation of the contemporary differences not only between the two communities under investigation but also in explaining differences between the two study communities and 'black spots' elsewhere in South Africa. 'Shack farming', for example, the practice wherein landowners lease plots to tenants for money rent, while prevalent in many areas of the country, has not occurred in Mgwali or Lesseyton. Clearly, the economically stagnant Eastern Cape does not offer the accumulation opportunities available to 'black spot' landowners in the vicinity of Durban (McCarthy, 1988; Jenkins et al., 1986).

However, in other areas, there are clear and very interesting parallels between the dynamics of the study communities and the broader South African land crisis. The attitudes of the landowners in Mgwali and Lesseyton appear very similar to those expressed by residents of white PWV suburbs being encroached upon by squatter settlement. Both display the 'militant conservatism' (Harris and Hamnett, 1987, p. 175) associated with neighbourhood consciousness and are essentially restorationist rather than revolutionary in orientation. Indeed, the Lesseyton landowners' demand to be granted the powers of removal extended to white farmers demonstrates that this group is well aware of the parallels between their situation and that of white landowners.

The complex relation between land, property and power, so stark in the study communities, appears to be gaining recognition at the national level. Claassens' (1991) theme that 'the right to property is the only classical human right which deals with the material world . . . [and therefore] is the only right which impinges directly on the material rights of others' (p. 11) indicates the potency of land rights as a source of power over others. People need space in which to live and if they 'have no rights to land and all the land around them is privately owned, they are forced by necessity to occupy or squat on land which belongs to someone else' (ibid.). In contemporary South Africa, the overwhelming need to find a place to live is forcing people into subordinate, and often marginal, positions 'under' landlords. The institutionalized bantustan circumstance where the power of the chiefs derives from land, is being reproduced informally elsewhere in South Africa with land remaining a potent instrument of patronage and control. Cross's (1991a) conclusion that 'land is . . . a social and political resource in the strict sense as well as a capital and subsistence asset' (p. 63) is as true of informal settlement in the former 'white' Republic as it is in the bantustans.

Such a conclusion has important implications for the broader debate on post-apartheid land tenure. First, it broadens the land-tenure question to encompass much more than the production issues which have dominated the debate thus far (Louw, 1988; Tapson, 1988). The findings from Mgwali and Lesseyton clearly indicate that land tenure informs societal relations far removed from production. Indeed, land ownership in the communities has remained critical to societal differentiation in the absence of production for there has been little if any agriculture in either community in the eighties.

Second, it changes the terms of reference of the land tenure debate. While agreeing with Cross (1985) that land tenure problems are not a major contributor to the stagnation of black agriculture in South Africa, a circumstance which is more accurately a product of other factors such as high input costs, inadequate rural transport and supply infrastructures and a dearth of agricultural and management skills and expertise, we are required to pursue the implication of this to its logical conclusion. It is apparent from the evidence above that local demands for land tenure change and specific tenure types do not simply and exclusively reflect production issues. The consistent demands for land ownership encountered in Mgwali and Lesseyton were prompted by a complex of political, cultural and social considerations, often far removed from production. This implies that, *ceteris paribus*, the extension of land ownership to all residents is unlikely to have any significant impact upon productivity, as interviewees were not demanding tenure changes for production-based reasons. A further corollary is that tenure changes which prompt production increase, but do not address broader societal issues, will not necessarily be well received in Mgwali or Lesseyton.

The last point draws us to an uncomfortable conclusion. To fulfil the basic subsistence needs of the Mgwali and Lesseyton communities necessitates a dramatic increase in local production. To achieve this first step on the path to recovery, significant changes are required in the use and control of the principal local means of production, the land resource. As noted, however, such changes would be poorly received, if not actively resisted, by the communities. Community demands, albeit expressing the will of the majority, and social justice prove irreconcilable if the most needy are actively contributing to their own marginalization. This raises the question as to which are more legitimate, local democratic demands or the abstract ideals expressed in a constitution?

These conclusions also have significant implications for the debate about land redistribution in a post-apartheid South Africa. In many ways the present is a fortuitous time for the government to proceed with land redistribution. With recent droughts in southern Africa driving down farm land prices and turmoil in international currency markets driving up the price of gold, large-scale land purchase by the state is more affordable now than for some years. However, as has been demonstrated, land is much more than an economic commodity, and its redistribution would have a significant impact on the political geography of contemporary South Africa. With substantial power shifts in the offing, the resistance to land redistribution at both the national and the local level is likely to be considerable, not merely because of the 'administrative tangle' (Evans, 1988, p. 136) involved but also as a result of vested interests not wishing to relinquish their power base in land. As is evident from studies elsewhere in South Africa (Stavrou and Crouch, 1989), such power struggles may assume political labels seemingly removed from the land, but this should not mask the fact that land issues underlie the conflict.

It is also apparent that assumptions about land issues and land relations being made in the political arena need to be broadened to encompass far more than production considerations. This requires a fundamental reappraisal of land politics that reaches far beyond the vision provided by a literal interpretation of the Freedom Charter (1955). To share the land 'among those who work it' (ANC, 1992, p. 1) would not only divide the country among a small minority of its population (Mabin, 1988) but would also act to perpetuate the view that land is essentially a means of production. Recent pronouncements by the ANC Land Commission seem to go some way towards recognizing this shortcoming. The Commission recognizes that at the national level, the 'return of our land is a necessary part of the return of our political power' (ANC, 1992, p. 2) but also that 'if a radical land reform does not take place we can expect an outburst of uncontrollable political anger' (p. 7) at the local level. However, the con- temporary circumstance wherein many blacks are keen to maintain both an urban property and a 'rural sheet anchor' (Moller, 1988) has already blocked several radical land reform options such as proposals to limit land ownership to one parcel of land per family unit.

Land occupies a unique position in South African society and its psyche. Both extremes of the political spectrum use land issues and imagery to fire up support for their cause. As Zulu (1988) notes, this is because land has become the primary 'medium through which the relations of exploitation and domination on the one side, and power and powerlessness on the other, are expressed' (p. 42). Any reform package which ignores the land issue can at best be peripheral, for an appreciation of the relationship between land, property and power is critical to the construction of any meaningful settlement to the current South African crisis. In this regard, it must be noted that reforms which concentrate on national divisions between black and white while ignoring local divisions between black and black will ultimately fail. At the community level, differentiation in land relations constitutes an obstacle which bars the road to social justice as surely as the legislative pillars of apartheid. This was certainly found to be the case in Mgwali and Lesseyton.

REFERENCES

ANC (1992) *Discussing the Land Issue: a Discussion Document for ANC Land Commissions and Branches*, ANC Land Commission, Johannesburg.

Bellman, H. (1927) *The Building Society Movement*, Methuen, London.

Bond, P. (1990) *The Urbanisation of Capital through Housing Finance in South Africa*, research paper, University of Zimbabwe, Harare.

Bromley, M. and Cernea, M. (1989) *The Management of Common Property Natural Resources: Some Conceptual and Operational Fallacies*, World Bank Discussion Papers, no. 57, The World Bank, Washington, DC.

Bundy, C. (1979) *The Rise and Fall of the African Peasantry*, Heinemann, London.

Claassens, A. (1991) *Who Owns South Africa: Can the Repeal of the Land Acts Deracialise Land Ownership in South Africa?*, Occasional Paper 11, Centre for Applied Legal Studies, University of the Witwatersrand, Johannesburg.

Cohen, R. (1970) 'Traditional society in Africa' in J. Paden and E. Soja (eds), *The African Experience*, vol. 1, Heinemann, London, 37–60.

Cokwana, M. (1988) 'A close look at tenure in Ciskei' in Cross and Haines, *Towards Freehold*, 305–14.

Cross, C.R. (1985) *If You Have No Land You are not a Man at All: Social Thought and the Mechanics of Land Tenure in Modern KwaZulu*, research paper, Centre for Applied Social Studies, University of Natal, Durban.

Cross, C.R. and Haines, R.J. (1988) *Towards Freehold: Options for Land and Development in South Africa's Black Rural Areas*, Juta & Co., Johannesburg.

Cross, C.R. (1991a) 'Informal tenures against the state: land holding systems in African rural areas', in M. de Klerk (ed.), *A Harvest of Discontent: the Land Question in South Africa*, IDASA, Cape Town, 63–98.

Cross, C.R. (1991b) *Structuring Land Tenure in a New Society: the Property Rights Question in Rural African Land Systems*, research paper, Rural–Urban Studies Unit, Centre for Social and Development Studies, University of Natal, Durban.

Crush, J. (1991) 'The discourse of progressive human geography', *Progress in Human Geography*, 15(4), 395–414.

Evans, R. (1988) 'Black land in white Natal: how to set up an administrative impasse' in Cross and Haines, *Towards Freehold*, 129–36.

Ferguson, J. (1990) *The Anti-politics Machine: 'Development', Depoliticisation and Bureaucratic State Power in Lesotho*, David Philip, Cape Town.

Haines, R.J. and Tapscott, C.P. (1988) 'The silence of poverty: tribal administration and development in rural Transkei' in Cross and Haines, *Towards Freehold*, 164–73.

Harris, R. and Hamnett, C. (1987) 'The myth of the promised land', *Annals of the Association of American Geographers*, 77 (2), 173–90.

Harvey, D. (1985) *Consciousness and the Urban Experience*, Johns Hopkins University Press, Baltimore.

Jenkins, D.P., Scogings, D.A., Margeot, H., Fourie, C. and Perkin, P. (1986) *An Investigation of the Emerging Patterns of Zulu Land Tenure and their Implications for the Establishment of Effective Land Information and Administration Systems as a Base for Development*, Occasional Paper, Department of Surveying and Mapping, University of Natal, Durban.

Louw, L. (1988) 'Black tenure versus white tenure in South Africa: the impact on development' in Cross and Haines, *Towards Freehold*, 294–301.

Mabin, A. (1988) 'Land ownership and the prospects for land reform in the Transvaal' in Cross and Haines, *Towards Freehold*, 137–45.

McCarthy, J. (1988) 'African land tenure as a popular issue and power relation in South African urban and peri-urban areas' in Cross and Haines, *Towards Freehold*, 294–301.

Moller, V. (1988) 'Some thoughts on black urbanisation after the abolition of influx control' in Cross and Haines, *Towards Freehold*, 146–57.

Slater, H. (1980) 'The changing pattern of economic relationships in rural Natal, 1838–1914' in S. Marks and A. Atmore (eds), *Economy and Society in Pre-industrial South Africa*, Longman, London, 148–70.

Sparks, A. (1990) *The Mind of South Africa*, Heinemann, London.

Stavrou, S. and Crouch, A. (1989) *Violence on the periphery: Molweni*, published paper, no. 4, Centre for Social and Development Studies, University of Natal, Durban.

Swalens, P. (1990) 'Provision of land: the key to peace', *The New African*, Urbanisation Supplement, **12**, September.

Tapson, D.R. (1988) 'Freehold vs leasehold in the homelands' in Cross and Haines, *Towards Freehold*, 329–36.

Tuan, Y.F. (1974) 'Space and place: a humanistic perspective', *Progress in Geography*, **6**, 211–52.

Zulu, P. (1988) 'The inadequacy of reform: land and the freedom charter' in Cross and Haines, *Towards Freehold*, 42–9.

5 Shades of 'Green' and 'Brown': Environmental Issues in South Africa

COLEEN H. VOGEL AND JAMES H. DRUMMOND

INTRODUCTION

The nineties have come to be known as the 'green decade'. From Rio to Johannesburg, environmental groups, politicians and citizens are talking and thinking 'green'. Issues including bio-diversity, the enhanced greenhouse effect, ozone depletion and the threatened destruction of forests are high on the international environmental agenda. Local 'green' debates tend to be focused more around rural and urban environmental problems arising from the political economy of apartheid (Downing, 1990; Cock and Koch, 1991; Ramphele and McDowell, 1991; Hallowes, 1993). This is not to say that the more traditional 'green' issues are not important in the South African context. The recent interest in the kaolin mining of Chapman's Peak in Cape Town and the controversy surrounding St. Lucia, Natal have focused attention on areas of outstanding natural beauty: '. . . the country's nature conservation state—the national parks, reserves, botanical gardens, are among the best managed conservation areas in the world' (Huntley, 1990, p. 95). Coexisting with this proud environmental heritage, however, is a suite of urban 'brown' environmental problems (Leitman *et al.*, 1992) such as poor quality water, inadequate sanitation, waste removal, air pollution and land degradation.

The argument advanced here is that 'green' issues in the country cannot be separated from the 'brown' issues which have their origins in the political and socio-economic policies of the apartheid state. Several of these, including pressure on land and inadequate provision of facilities, both in urban and rural areas, have roots in broader national historical and political planning, including 'betterment' planning, which needs to be remembered when examining South African environmental problems (Beinart, 1984, 1989; De Wet, 1987; Wilson and Ramphele, 1989; Ramphele and McDowell, 1991; Drummond and Manson, 1993; Vogel and Drummond, 1993). It is against this backcloth that reference will be made to case studies in both rural and urban contexts. Future resource management options to safeguard and ensure sustainable use of South African environments will also be suggested.

The Geography of Change in South Africa, edited by Anthony Lemon.
© 1995 by the Editor and Contributors. Published in 1995 by John Wiley & Sons Ltd.

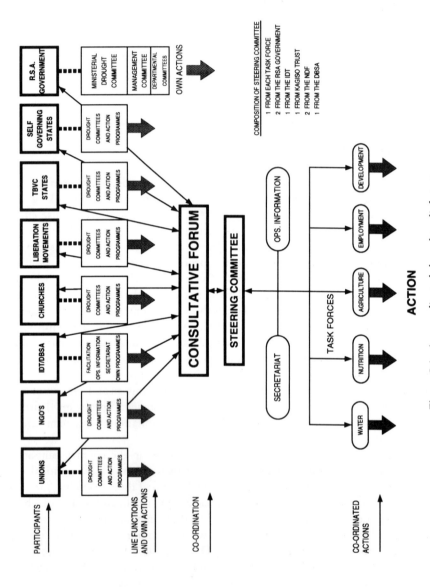

Figure 5.1. A co-ordinated drought relief strategy

LAND DEGRADATION

South African climates are characterized by an erratic rainfall, hot summers and dry winters with winter rainfall usually restricted to the south-western areas of the country. As a result, much of the country, particularly towards the west, is dry and receives less than 500 mm of rain. Several of these areas, moreover, have high population densities with environmental problems which are aggravated by poor infrastructure and provision of water (Bekker et al., 1992). Approximately 53 per cent of the population do not have access to clean water (Khan, 1993).

Several rural environmental problems abound, including land degradation and soil erosion, which are often exacerbated during drought periods. Soil erosion is particularly prevalent in the Eastern Cape, Lesotho and Natal. Some writers (Boucher and Weaver, 1991; Watson, 1991; Garland and Broderick, 1992; Garland and Pile, 1993) ascribe these soil erosion problems mainly to the management of local environments and to the normal physical changes in the geomorphology of the landscape, whereas others (Wilson and Ramphele, 1989; Corbett, 1990; Ramphele and McDowell, 1991) attribute soil erosion and excessive land degradation more to overpopulation of the bantustans, caused by the socio-political strategies of the past. In 1980, for example, in the rural Transvaal, excluding the bantustans, the population density was $11/\text{km}^2$ whereas in the Transvaal as a whole, the rural density was $47/\text{km}^2$ (Wilson, 1991). Skewed distributions of population, brought about by resettlement, forced removals and the creation of the bantustans, have therefore exerted great pressure on the land

Figure 5.2. Recent View of Squatter Settlement (Alexandria) and Jukskei River (Photograph: S. Mhlungu)

and have exposed such negative environmental ills as soil erosion and deforestation. These environmental problems are not spatially uniform and marked
landscape variations occur (Drummond, 1992; Vogel, 1994b; Dean et al., 1994).

The occurrence of frequent drought and unsustainable agriculture in certain
areas has served to lay bare a number of these rural environmental problems.
Sixty-five per cent of South Africa is unsuitable for dryland farming (Huntley et
al., 1989), having an annual rainfall of less than the agricultural minimum for
successful crop farming. Despite the unfavourable physical resources of the
country, state subsidy policies in the past meant that farming in marginal areas
was possible. Food price increases above the average rate of inflation, increasing
producer prices, the restructuring of agriculture and the withdrawal of blanket
drought assistance to farmers have all exacerbated the impacts of drought on the
commercial farming community (Vogel, 1994b). Since the mid-1980s, the nexus
of drought, escalating debts and increasing agricultural producer prices meant
that farming in marginal agricultural areas was no longer viable, and specific
policies, designed to discourage the farming of marginal lands, were introduced
(Vogel, 1994a).

The social dimensions of drought (Figure 5.1) have been given a high profile by
the National Consultative Forum on Drought (established at the height of the
1992 drought). Water, nutrition and early warning task forces have repeatedly
drawn attention to the fact that the impacts of drought are severe, not only
because of a shortage of rainfall but because of mismanagement and a lack of
development, particularly in black rural areas. Reports have been made of
infrastructure that has not been maintained, large distances that have to be
traversed by rural communities to reach water and the lack of awareness of
community drought-coping mechanisms (Abrams et al., 1992). Regional examinations of drought impacts in such areas as Bophuthatswana have also indicated
that the lack of adequate planning, particularly pro-active drought management,
coupled to development (Vogel, 1994b; Vogel and Drummond, 1993), will
continue to frustrate well-intentioned drought relief.

Deforestation and bush encroachment are additional environmental problems
in the country (Milton and Bond, 1986). In most rural settlements wood is an
important energy resource. The excess of wood demand over supply contributes to
the increased degradation of this valuable resource. Each year roughly eight
million tons of wood are burned in the country as domestic fuel (Cock and Koch,
1991, pp. 94–109). At a regional level the scale of wood use is also high with
estimates in Gazankulu, for example, indicating that wood supplies between 58 per
cent and 98 per cent of household energy requirements. Depletion of this resource
is unfortunately rapid, and in KwaZulu, of the 250 forests proclaimed in terms
of the 1936 Land Act, a mere one-fifth are still intact today (Cock and Koch).

One of the areas where struggles around the rural environment have been
reflected is in the dilemma between development and conservation. Conservation
in Africa has often been shown to be at odds with rural development (Beinart,

Figure 5.3. The Impact of Drought in Bophuthatswana (Photograph: J. Alfers)

1984, 1989) not least in South Africa (e.g. Carruthers, 1989; Ramphele and McDowell, 1991, pp. 155–67; Brooks, 1992; Munnik *et al.*, 1993): 'Until recently people from the villages and townships around our game reserves have been victims rather than beneficiaries of conservation' (Munnik *et al.*, 1993, p. 1). Progress in this important area is being made as communication between rural communities and various conservation bodies improves, with greater involvement and management of such areas accruing to the community, for example the Richtersveld National Park in the Northern Cape and Pilanesberg in Bophuthatswana (Ramphele and McDowell, 1991, pp. 155–67; Cock and Koch, 1991, pp. 112–28; Munnik *et al.*, 1993).

URBAN ENVIRONMENTAL PROBLEMS

Having illustrated some of the critical rural environmental matters, we now turn to examining some of the pressing, but poorly documented, urban environmental problems. Increased urbanization, associated with complex rural migration patterns and natural growth in metropolitan areas, is increasing the stress placed on the urban environment (Ramphele and McDowell, 1991, pp. 91–102; Posnik and Vogel, 1992). Urban 'brown' environmental problems such as inadequate waste management, poor water quality, accidents linked to crowding and congestion in squatter areas are now assuming a higher place on the local environmental agenda. Several of these problems in South Africa owe their origins to state policies which have created townships which contain islands of extreme poverty and squalor.

A sizeable proportion of the urban environmental ills besetting metropolitan South Africa is thus linked to the lack of housing, land and the provision of essential services. The Palmer study (Palmer Development Group, 1992) for example, has shown that approximately 31 per cent of the estimated 24.5 million people living in urban areas do not have access to adequate sanitation. The magnitude of the problem may be far worse, as the figures are probably under-estimates. In 1987 for example, an estimated 120 000 people lived in Alexandra, an area of approximately 5 km^2, with 9240 people housed in hostels (Mashabela, 1988). It is not surprising, therefore, that there was and continues to be a lack of housing (an average of one house for every 10 people) and adequate service provision. Bucket sewerage, poor and inadequate waste removal and blocked stormwater drains cannot compete with the growing demands of the population (Posnik and Vogel, 1992).

These problems are prevalent in several other urban areas. In Bekkersdal for example, residents in 1987 had no sewerage reticulation system (Mashabela, 1988), and in Dobsonville and Diepmeadow (area of Soweto) in the early 1990s, only 65 per cent and 80 per cent of the population, respectively, had water-borne sanitation (Palmer Development Group, 1992). Several attempts to upgrade these urban townships have taken place, but shortages of essential services still exist (Posnik and Vogel, 1992). Increased population pressures in informal settlements, where data on environmental quality of life are poor, only serve to compound these urban environmental problems.

ENVIRONMENTAL PROBLEMS WITHOUT BOUNDARIES

The distinction between rural and urban environmental problems is often diffi-cult to make. The boundary between regions is often blurred and environmental problems arising in one area, such as water and air pollution, are frequently imported into neighbouring places. Although South Africa's calculated propor-tion of the contribution to the enhanced greenhouse effect amounts to a mere 2 per cent of the world's total (Republic of South Africa, 1991), local levels of pollution are still cause for concern, particularly in terms of human health. In 1987 for example, approximately 332 million tons of carbon dioxide emissions were measured originating in South Africa, with 250 million tons arising from the energy industry, 12 million tons from general industry, eight million tons from households, two million from steel and alloy and 60 million from motor vehicles (Republic of South Africa, 1991).

Power stations contribute significantly to the pollution problems of the country. Some 83 per cent of South Africa's electricity is generated by coal combustion, containing about 1.2 per cent of sulphur and a higher percentage of ash (between 15 and 50 per cent) (CSIR, 1991). Pollution conditions in the Eastern Transvaal Highveld (ETH), an area of approximately 30 000 km^2, are augmented by the stable character of the atmosphere that impedes pollution dispersal and also by the high incidence of industrial and agricultural activities in the area (Tyson et al., 1987).

The physical implications of such pollution include the threat of acid rain. Measures of the pH of normal rainfall indicate mild acidity (5.6). In the ETH direct measures of rainfall acidity range from 3.9 to 4.6 (Fuggle and Rabie, 1992, pp. 417–55). Despite the low wet and dry nitrates values in the ETH, which are generally lower than industrial sites elsewhere in the world because of a low contribution from the transport sector (Fuggle and Rabie), the pollution from these sources merits close monitoring.

Air pollution is also a ubiquitous problem in urban environments. The bulk of residential pollution in South Africa is a consequence of domestic coal consumption for cooking and space heating (Fuggle and Rabie). Pollution conditions in certain townships (such as those in the Vaal Triangle and Soweto) have been and are being monitored (for example Rorich, 1988; Turner et al., 1986). In the least polluted parts of the country, such as the north-western Transvaal, monitored mean annual sulphur concentrations are about seven micrograms per cubic metre and hourly means only exceed 29 micrograms per cubic metre for 1 per cent of the time. In Soweto, however, annual mean concentrations have been shown to be as high as 60 micrograms per cubic metre. Although domestic emissions only account for 3 per cent of sulphur dioxide and 24 per cent of all particulates emitted in South Africa, the impacts remain severe, particularly when it is noted that concentrations, such as those in Soweto, are experienced by millions of township residents (CSIR Environmental Services, 1992).

Asbestos, another air-borne pollutant, can result in severe environmental health problems and unfortunately also knows no territorial delimitation. South Africa produces a high percentage of the world's crocidolite (blue asbestos) mined in parts of the northern Transvaal and northern Cape. A serious environmental issue emerging from asbestos exposure is the fatalities due to cancer of the bronchus, stomach and lungs: 'Death rates in the general population living in certain mining areas of South Africa have been shown to be higher than in control areas' (CSIR Environmental Services, 1992, p. 70). Asbestos diseases are not only confined to the mining areas, for contamination can occur in the shipment and wind erosion of the mineral fibres (Cock and Koch, 1991, pp. 33–43).

WATER: ACCESS AND QUALITY

Water availability is often a more insidious problem than atmospheric pollution. Potable water is frequently inaccessible, and it has been estimated that

Figure 5.4. Poor Infrastructure and Water Provision in Black Rural Areas (Photograph: J. Alfers)

approximately 33 per cent of the urban population of the country has minimal access to sanitation and that 18 per cent have minimal access to water (Water and Sanitation 2000 Working Group, 1991). Freshwater pollution, originating mainly from industrial, domestic and agricultural sources, amounts to some 780 000 tons annually with storm-water run-off contributing 194 000 tons of pollutants to marine environments annually (CSIR, 1991). Poor water quality can result in numerous diseases including gastro-intestinal infections as well as being harmful to other components of the local ecology. In South Africa, diarrhoea accounts for over 10 000 deaths annually with the majority being children under the age of five (Yach *et al.*, 1989). Rural areas, because of inadequate water supply, account for the highest number of these cases.

The water quality in rivers traversing built environments, such as Johannesburg, is also known to be environmentally degraded. Water courses in the area are often used as dumping grounds for neighbouring industry, squatter communities abutting the rivers and others in the area. It is not surprising therefore that incidents of toxic pollution occurrences in these rivers are often given press coverage (for example, *Sunday Times*, Johannesburg, 25 Oct 1991). Pollution of urban water has also been noted in other areas. Results of tests taken on samples by laboratories in 1991 and 1992 for the Jukskei River in Johannesburg thus showed that Coliform Bacteria counts, Faecal Bacteria and *E Coli* type 1 were above acceptable limits (Posnik and Vogel, 1992). Pollutants, including nitrates, phosphates and ammonia have also been shown to be above acceptable limits in other Johannesburg rivers such as the Braamfontein Spruit. Certain of these pollutants, however, do appear to be filtered out by dams in the area (Gaffin *et al.*, 1989).

FUTURE ENVIRONMENTS

Despite the world-acclaimed ecological attractions of the country, large tracts of South Africa have been shown here to be beset with urban or rural environmental problems. Notwithstanding these drawbacks, South African environmental problems are receiving increasing attention from the government and various institutions (Council for the Environment, 1989). In addition, non-government organizations, namely the Built Environment Support Group, Earthlife Africa and the Group for Environmental Monitoring (Hallowes, 1993), have made substantial progress in calling for increased monitoring of environmental issues in the country and pressing for accountability of those interfering with the environment in an unsustainable manner. Integrated Environmental Management (IEM), despite criticisms of not embracing the wider environmental concerns of the country (Wulfsohn, 1991; Coovadia *et al.*, 1993; Quinlan, 1993), has laid the foundations for an interactive and holistic framework for understanding environmental problems and improving environmental planning. One example of the need for greater interaction between the state and communities is the controversial mining of St. Lucia which has highlighted both the need for local advocacy and future state and industrial accountability.

Developmental and environmental interests have been locked in a battle over the St. Lucia greater wetlands area in Natal (Figure 5.5). Richards Bay Minerals (RBM) applied for the rights to mine the area in 1989, but before permitting them to go ahead, the government called for an Environmental Impact Assessment (EIA) to be completed. The area holds both great scenic and natural wealth (wetlands and lakes) and also a store of mineral wealth (rutile, ilmenite and zircon) that is contained in the shallow capping on the dunes (CSIR Environmental Services, 1993). Both the nature conservation lobbyists and the mining company have particular interests in the area. The former wish to preserve it in the interests of conservation of a unique wetland and for ecotourism. The latter primarily seeks to be able to mine the area in conjunction with '. . . conservation and tourism activities where feasible' (CSIR Environmental Services, 1993, p. 11).

The EIA on St. Lucia has unearthed several issues that some argue have been ignored. These issues include how the proposed mining of the dunes has, and will, impinge on the lives of local residents (*The Star*, Johannesburg, 15 May 1993). While those undertaking the EIA of St. Lucia have tried to follow an integrated environmental management approach (Fuggle and Rabie, 1992), that includes a procedure whereby public participation is sought at all levels, there are

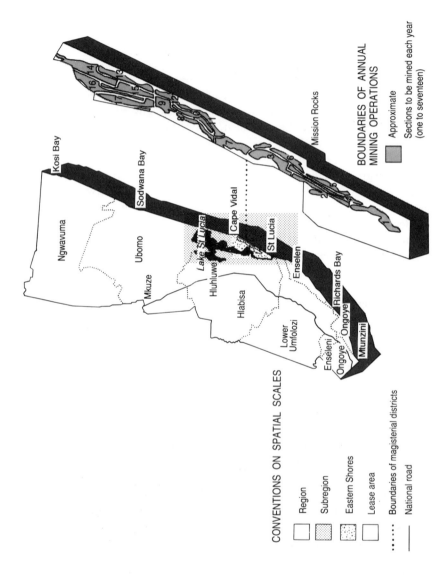

Figure 5.5. Mining in the St. Lucia wetlands

CONVENTIONS ON SPATIAL SCALES

- Region
- Subregion
- Eastern Shores
- Lease area
- Boundaries of magisterial districts
- National road

BOUNDARIES OF ANNUAL
MINING OPERATIONS

- Approximate
- Sections to be mined each year
  (one to seventeen)

Kosi Bay

Ngwavuma

Ubomo

Sodwana Bay

Mkuze

Lake St Lucia

Cape Vidal

Hluhluwe

St Lucia

Enselen

Hlabisa

Lower
Umfolozi

Richards Bay

Enseleni

Ongoye

Ongoye

Mtunzini

Mission Rocks

those who would argue that this has not taken place. The rights of the local people were eliminated in 1973 through forced removal, allowing RBM, the Natal Parks Board and other green groups to wrangle over the land. Such issues, compounded by grievances that include those of mineworkers attached to RBM and vocalized through their union (the National Union of Mineworkers), echo some of the real environmental issues besetting the country, namely the allocation of land and the equitable distribution of resources. Recent strong recommendations to government, who have yet to make a decision on the issue, have proposed that mining not be permitted and that the land issue be carefully examined.

Several emerging political groups have also begun to respond to issues of the environment such as the environmental implications of pollution, sustainable development and the land issue (Desai, 1991; Sisulu and Sangweni, 1991). More specific endeavours which could improve the local environment of the future include active research into programmes on sustainable agriculture, engendering an ethnoecological approach such as agroforestry, assisting in the regrowth of forests (Milton and Bond, 1986; Banks *et al.*, 1993) and investing in appropriate environmental education.

Non-government organization and community efforts are endeavouring to involve communities in sustainable environmental activities (South African Assembly on Women and the Environment, 1993). Intensive research of urban environmental problems must be undertaken and these must be located firmly on future planning agendas. The continued extension of the interaction between all stakeholders in environmental debates must also be ensured.

CONCLUSION

Green environmental issues such as ozone depletion and the enhanced greenhouse effect are pressing environmental concerns. Poverty, however, that manifests itself in the form of 'brown' environmental issues, including rural and urban land degradation and poor quality of life, is one of the more immediate environmental concerns facing South Africa. Development, based on sustainable environmental practices, is going to become increasingly important in South Africa. Alternative approaches that enable local populations either to revive past capacity, or to create new ways of managing the environment, will have to be actively pursued. Environmental education, both to impart environmental management techniques and to enable people to recapture their ecological literacy, is essential when trying to change environmental hues from 'dirty' browns to 'brilliant' greens: 'Hereafter [the environmental movement] will either be lugging its legacies along as it tilts at windmills. Or it will be seeking out its commonalities with other, often broader, struggles for democracy and justice' (Marais, 1993, p. 35). The hope is that the latter will prevail.

ACKNOWLEDGEMENT

James Drummond is grateful for financial assistance from the University of the North West to support research which contributed to this chapter.

REFERENCES

Abrams, L., Short, R. and Evans, J. (1992) *Root Causes and Relief Restraint Report*, Consultative Forum on Drought, Johannesburg.

Banks, J., Griffin, N., Mavradonis, J., Shackleton, C. and Shackleton, S. (1993) 'Planting trees, who does, who doesn't and why', *New Ground*, 13, 18–19.

Beinart, W. (1984) 'Soil erosion, conservationism and ideas about development: a southern African exploration, 1900–1960', *Journal of Southern African Studies*, 11, 52–83.

Beinart, W. (1989) 'Introduction: the politics of colonial conservation', *Journal of Southern African Studies*, 5, 143–62.

Bekker, S., Cross, C. and Bromberger, N. (1992) 'The wretched of the earth', *Indicator South Africa*, 9(4), 53–60.

Boucher, K. and Weaver, A. (1991) 'Sediment yield in South Africa—a preliminary geographical analysis', *GeoJournal*, 23 (1), 7–17.

Brooks, S. (1992) 'The environment in history: new themes for South African geography' in C. Rogerson and J. McCarthy (eds), *Geography in a changing South Africa: Progress and Prospects*, Oxford University Press, Cape Town, 158–72.

Carruthers, J. (1989) 'Creating a national park, 1910–1926', *Journal of Southern African Studies*, 15 (2), 188–216.

Cock, J. and Koch, E. (eds) (1991) *Going Green: People, Politics and the Environment in South Africa*, Oxford University Press, Cape Town.

Coovadia, Y., Dominik, T., Walton, B. and Wulfsohn, T. (1993) 'How green is my urban development NGO: sustainable development responses from Built Environmental Support Group' in D. Hallowes (ed.), *Hidden Faces: Environment, Development, Justice: South Africa and the Global Context*, Earthlife Africa, Scotsville, 156–84.

Corbett, B. (1990) Development for people and plant, *Rotating the Cube, Environmental Strategies for the 1990s, an Indicator SA Focus*, Department of Geographical and Environmental Studies and Indicator, University of Natal, Durban.

Council for the Environment (1989) *Integrated Environmental Management in South Africa*, Joan Lotter, Pretoria.

CSIR (1991) *The Situation of Waste Management and Pollution Control in South Africa*, Report to the Department of Environmental Affairs by the CSIR Programme for the Environment, Report CPE1/91.

CSIR Environmental Services (1992) *Building the Foundation for Sustainable Development in South Africa*, National Report to the United Nations Conference on Environment and Development (UNCED), Rio.

CSIR Environmental Services (1993) *Environmental Impact Assessment, Eastern Shores of Lake St. Lucia*, Summary Report, January.

Dean W.R.S., Hoffman, M.T., Meadows, M.E. and Milton, S.J. (1994) 'Desertification in the semi-arid Karoo, South Africa: review and reassessment', *Journal of Arid Environments*, in press.

Desai, B. (1991) 'An environmental policy for the Pan Africanist Congress of Azania', *History in the Making: Documents Reflecting a Changing South Africa*, 2, 46–9.

De Wet, C.J. (1987) 'Betterment planning in South Africa: some thoughts on its history, feasibility, and wider policy implication', *Journal of Contemporary African Studies*, 6 (1–2), 85–122.

Downing, A.B. (1990) *Apartheid's Environmental Toll*, Worldwatch Paper, 95, World Watch Institute, Washington, DC.

Drummond, J.H. (1992) *Changing Patterns of Land Use and Agricultural Production in Dinokana Village, Bophuthatswana*, Unpub. MA dissertation, University of the Witwatersrand, Johannesburg.

Drummond, J.H. and Manson, A.H. (1993) 'The rise and demise of African agricultural production in Dinokana village, Bophuthatswana', *Canadian Journal of African Studies*, 27 (3), 462–79.

Fuggle, R.F. and Rabie, M.A. (1992) *Environmental Management in South Africa*, Juta, Cape Town.

Gaffin, A., Kaplan, M. and Simon, C. (1989) 'Chemical pollution levels of dam systems', *Spectrum*, 27, 16–18.

Garland, G.G. and Broderick, K. (1992) 'Changes in the extent of erosion in the Tugela catchment, 1944–1981', *South African Geographical Journal*, 74, 45–8.

Garland, G.G. and Pile, K. (1993) 'Bodies on the move: Geomorphology and development at Cornfields, Natal', paper presented at Society for Geography Conference, 18 July, Port Elizabeth.

Hallowes, D. (ed.) (1993) *Hidden Faces: Environment, Development, Justice: South Africa and the Global Context*, Earthlife Africa, Scotsville.

Huntley, B. (1990) 'Conservation 2000—opportunities', paper presented at the Conservation 2000 Symposium, 18 September, Pretoria.

Huntley, B., Siegfried, R. and Sunter, C. (1989) *South African Environments into the 21st Century*, Human and Rousseau Tafelberg, Cape Town.

Kahn, F. (1993) 'Rural development: the case of Riverlands', *Earth Year*, 4, 10–11.

Leitman, J., Bartone, C. and Bernstein, J. (1992) 'Environmental management and urban development: issues and options for Third World cities', *Environment and Urbanisation*, 4, 131–40.

Marais, H. (1993) 'When green turns to white', *Work in Progress*, 89, 34–6.

Mashabela, H. (1988) *Townships of the PWV*, South African Institute of Race Relations, Johannesburg.

Milton, S.J. and Bond, C. (1986) 'Thorn trees and the quality of life in Msinga', *Social Dynamics*, 12, 64–76.

Munnik, V., Moloi, D. and Ngqobe, Z. (1993) 'People and parks', *New Ground*, 13, supplement.

NAMPO (National Maize Producers Organization) (1993) *Maize Production in the 1990's*, Dreyer, Bloemfontein.

Palmer Development Group in association with the University of Cape Town (1992) Urban sanitation evaluation regional profile, draft report.

Posnik, S. and Vogel, C. (1992) 'Low-income housing—the "brown" environmental issues', *Proceedings of the EPPIC '92 Conference on 'Poverty and the Environment'*, 28–9 September, Midrand, Johannesburg, 117–28.

Quinlan, T. (1993) 'South Africa's integrated environmental management policy at the crossroads of conservation and development', *Development Southern Africa*, 10, 230–8.

Ramphele, M. and McDowell, C. (1991) *Restoring the Land, Environment and Change in Post-apartheid South Africa*, Panos, London.

Republic of South Africa, President's Council Report (1991) *Report of the Three Com-*

mittees of the President's Council on a National Environmental Management System, Government Printer, Pretoria.

Rorich, R.P. (1988) Soweto Air Pollution in July to August 1984, Residential Air Pollution, National Clean Air Association, 1988.

Sisulu, M. and Sangweni, S. (1991) 'Future environment policy for a new South Africa', ANC discussion paper on environment policy for a new South Africa, History in the Making: Documents Reflecting a Changing South Africa, 2, 37–45.

South African Assembly on Women and the Environment (1992) 'Health and safety aspects of domestic fuels', Draft National Energy Council Report, Medical Research Council, Pretoria, 1992.

Turner, C.R., Annegarn, H.J., Rorich, R., Wells, B., Turner, S. and Page, K. (1986) 'Air pollution monitoring in Soweto', Proceedings of the International Conference on Air Pollution, CSIR, Pretoria, 87–109.

Tyson, P.D., Kruger, F.J. and Louw, C.W. (1987) Atmospheric Pollution and its Implications in the Eastern Transvaal Highveld, FRD, South African National Scientific Programmes Report, 150, CSIR, Pretoria, 1–67.

Viljoen, R. (1991) 'Going up in smoke', New Ground, 5, 10.

Vogel, C.H. (1994a) 'South Africa' in M.H. Glantz (ed.), Drought Follows the Plough, Cambridge University Press, 151–70.

Vogel, C.H. (1994b) 'Consequences of drought in Southern Africa, 1960–1992', unpublished Ph.D. thesis, University of the Witwatersrand, Johannesburg.

Vogel, C.H. and Drummond, J.H. (1993) 'Dimensions of drought: South African case studies', GeoJournal, 30 (1), 93–8.

Water and Sanitation 2000 Working Group (1991) Workshop on strategies for water supply and sanitation, Johannesburg.

Watson, H.K. (1991) A Comparative Study of Soil Erosion in the Umfolozi Game Reserve and Adjacent Kwa Zulu Area from 1937–1983, unpublished Ph.D. thesis, University of Durban-Westville.

Wilson, F. (1991) 'Land out of balance' in M. Ramphele (ed.), Restoring the Land, Environment and Change in Post-apartheid South Africa, Panos, London, 27–38.

Wilson, F. and Ramphele, M. (1989) Uprooting Poverty, the South African Challenge, David Philip, Cape Town.

Wulfsohn, T. (1991) 'Challenging the conventional approaches to natural resource management environmental planning in rural development: a regional political ecology analysis of Maputaland', Development Southern Africa, 8 (4), 495–508.

Yach, D., Strebel, P.M. and Joubert, G. (1989) 'The impact of diarrhoeal disease on childhood deaths in the RSA', South African Medical Journal, 76, 472–5.

6 Trends in Health and Health Care in South Africa

GARRETT NAGLE

INTRODUCTION: APPROACHES TO THE STUDY OF HEALTH AND HEALTH CARE

Health and health care issues can be tackled from many angles, ranging from the biomedical to the economic and the ethical, and each approach brings with it its own interests and bias. For example, the medical–biological approach is orientated more towards the individual, the anthropological or social-science approach is generally household-based and that of the economist is often theoretical. While medical interventions are generally at the individual level, other potential intervention strategies can operate in a number of alternative situations. This has created in many instances an interdisciplinary vacuum, since without discussion and cross-fertilization of ideas a true holistic approach cannot be achieved.

By contrast, it is argued that medical geography takes into account the multi-disciplinary nature, causes and effects of disease and analyses potential solutions to the problems. A geographical approach can operate at a variety of temporal and spatial scales from the individual or community health survey to national health policy or global food programme. Thus it can offer a unique appraisal of the conditions which operate in a given location at a particular time. Moreover, it synthesizes and, in an accessible way, presents data from a variety of primary and secondary sources.

Two main geographical approaches to health and health care may be identified: disease ecology and the provision of health care services (Jones and Moon, 1987; Learmouth, 1988). These have often been isolated from each other and have, at times, resembled the physical–human dichotomy within geography. However, some recent studies have adopted a more interdisciplinary approach and have linked the two during regional investigations into patterns of health and health care (Nagle, 1992).

The first approach, disease ecology or epidemiology, is mainly concerned with the distribution of diseases over space and an understanding of the processes which bring about such a distribution. For example, as a country develops there will frequently be a change in the disease pattern from largely infectious, contagious diseases such as TB, cholera and measles to mostly degenerative ones, characterized by cardio- and cerebro-vascular diseases and

The Geography of Change in South Africa, edited by Anthony Lemon.
© 1995 by the Editor and Contributors. Published in 1995 by John Wiley & Sons Ltd.

malignant neoplasms. This change has been referred to as the epidemiological transition.

The second approach has tended towards the analysis of the delivery and utility of health care. Many of these have involved the application of refined statistical techniques. In general, investigations have moved away from the location and numbers of facilities and personnel to more in-depth anthropological and sociological appraisals concerning access to health care and constraints imposed upon access.

DATA SOURCES

In South Africa, the majority of the relatively small number of human geographers concentrate on the problems of urbanization and demography, industrialization and development. Thus, there are few studies on health and health care by geographers. Consequently, the material required has to be extracted from official statistics, learned journals and a wealth of household surveys, many of which were carried out for the Second Carnegie Inquiry into Poverty and Development in Southern Africa, held in 1984.

Before discussing trends in health and health care in South Africa, it is important to consider the information that is available. Although health is defined as a state of complete physical, mental and social well-being, and not merely the absence of disease or infirmity, the data available are limited to morbidity (disease) and mortality. As yet there is no reliable systematic collection of statistics that covers all of South Africa: the so-called 'independent homelands' were omitted from official statistics for a long period of time and estimates of base population figures are unreliable, given population mobility and a lack of reliable population censuses. Likewise, the reliability of diagnosis is often questionable as for example with AIDS, influenza and TB. Moreover, not all areas are adequately monitored by the health services, and those in periurban areas, where there is a great deal of population mobility, are rarely covered well. Under-registration of deaths among blacks is well known, and secondary causes of death are rarely recorded on mortality data. Records that relate to morbidity are scarce and often inaccurate. Thus, for example, malnutrition as a cause of death or as an illness is rarely mentioned, although high rates of death and increased risk of infections are closely related to nutritional status (Nagle, 1992). Nevertheless, despite these limitations, mortality data can be used to provide an indication of the disease pattern and health status of a group.

A second source of data relates to published material such as in medical journals. However, much early literature submitted to the *South African Medical Journal*, one of the major sources of information, was inadequate, for example in the way it presented the nutritional status of blacks. The surveys undertaken tended to overemphasize ignorance and poor food habits as the causes of malnutrition, and many were more concerned with the biochemical rather than the social aspects of nutrition. Similarly, the South African Institute of Ra tions, in stressing the role of drought as a cause of malnutrition rather th lack of entitlement or command over resources, failed to identify the su mechanisms, as opposed to the physical triggers, which may precipitate disea: (SAIRR, 1970).

Owing to the rapid changes that are taking place in South Africa many studies date very quickly. Thus data collected for the Carnegie Conference are over a decade old and there have been fewer studies since the conference. Nevertheless, it is important to build on these studies, many of which are in-depth appraisals of health and access to health facilities throughout South Africa (for an overview see Wilson and Ramphele, 1989).

MORBIDITY AND MORTALITY IN SOUTH AFRICA

It is difficult to be certain of the health conditions of the population of South Africa before the early 20th century given the paucity of material available. However, the reports that exist suggest that although death rates were quite high, especially in times of famine, in general the physique, diet and health were good (Van der Kemp, 1804, 1812; Bryant, 1939). As in 19th-century Great Britain, industrialization in South Africa in the late 19th and early 20th centuries was associated with 'social' diseases, notably TB and venereal disease, the former subsequently becoming endemic in rural South Africa, largely as a result of returning African migrants (Gluckmann, 1944; Burroughs, 1958).

By the mid-20th century, the disease profile of South Africa tended to follow very clearly defined racial lines and, in general, this has persisted to the present although there have been recent changes. For whites the main diseases that cause morbidity and mortality are mostly degenerative such as cerebro-vascular diseases (strokes), cardiovascular disease (heart) and carcinoma (cancers). These account for over two-thirds of all deaths among whites, while infectious and parasitic diseases account for only two per cent. This is typically the pattern observable in a developed country. On the other hand, a very high proportion of the coloured and black population suffer high rates of infectious, contagious diseases such as TB, measles, gastro-enteritis and respiratory infections. By contrast, for blacks the rates for degenerative and infectious diseases are both approximately twenty per cent. Thus, in terms of the epidemiological transition model, South Africa's population is spread across the spectrum from degenerative diseases (mostly among whites) to contagious ones (mostly among blacks and coloureds). However, although not a strict dichotomy, the pattern was relatively clear. Today, however, urban blacks are increasingly taking on a pattern which mirrors that of whites, with high rates of strokes, heart diseases and cancers (Seedat, 1989; Steyn and Fourie, 1990; Steyn et al., 1990, 1991). For many households, the pressures borne by those involved in family break-ups, by single families, teenage parents, the unemployed and underemployed and those with few qualifications or skills have driven many to lifestyles which are 'self-destructive' through drugs, prostitution and alcoholism. Some recent studies have dealt with the increasing psychological disorders common among blacks (Turton and Chalmers, 1990; Vogelman, 1990). There has also been much attention given to the role of violence as a direct and indirect effect on the health of many South Africans some of which has taken place within the custody of the authorities.

Geographically, some areas are far worse off than others. In resettlement camps, such as Glenmore and Dimbaza in Ciskei, and periurban areas, the infectious type of disease is much more dominant, especially during times of rapid population influx, whether forced or voluntary. Such patterns can be largely attributed to the low incomes, inadequate diets, overcrowded housing, poor water and sanitation facilities which create a highly unstable social and economic environment. Although the apartheid policies which led to the underdevelopment of the homelands and those which caused forced removals have been dismantled, there are still many social and economic factors which prohibit many South Africans from achieving full health. These operate at a variety of spatial scales and include, among others, rapid population growth, recession, inflation and escalating violence.

CHILDHOOD MORBIDITY

The major types of illness experienced by coloured and black children are the infectious, communicable diseases such as measles, gastro-enteritis and respiratory diseases (Househam and Bowie, 1988; Tatley and Yach, 1988). Malnourished individuals have been shown to be at greater risk than well-nourished ones, especially with regard to gastro-enteritis, measles, TB, scabies and pneumonia (Nagle, 1992). In particular, the vast majority of those considered at risk of malnutrition are coloureds and blacks in periurban and rural areas. Recent studies also estimate high rates of malnutrition among blacks, ranging between 15 and 25 per cent. In some places, as much as 31 per cent of rural pre-school children were reported to be underweight (SAIRR, 1993). Indeed, the most recent reports indicate that rates of marasmus and kwashiorkor (chronic shortages of energy and protein respectively) have increased since the mid-1980s and, in the Cape Flats, there has been an increasing number of deaths from malnutrition (SAIRR, 1990, 1992). Some of these surveys may even underestimate the true extent of malnutrition, since they are based on creche- or school-attending children, and some exclude the populations of 'independent' homeland.

By contrast, white children generally experience a developed-world type disease pattern. Moreover, not only do the types of illnesses differ among the races but so do the frequency, severity and the long-term effects.

ADULT MORBIDITY

For black and coloured adults, a Third World type of disease pattern pre-dominates. The worsening recession has had an adverse effect on many diseases, such as measles and TB, as shown in Table 6.1.

Women, especially black women, constitute an extremely vulnerable popula-tion. They are at increased risk of morbidity and mortality given their low socio-economic standing, and many mothers are young, unmarried, malnourished and under tremendous stress. In a survey of women in Ciskei by the author, 77 per cent admitted to being nervous, tense and worried, 73 per cent were easily tired and suffered frequent headaches, while 69 per cent claimed to be tired all the time. Many referred to an inability to sleep, poor appetite and a general lack of emotional stability.

Reports suggest high rates of adult malnutrition, especially among blacks and coloureds. Wasting (underweight) is quite common among men and obesity is relatively frequent among women. However, in some of the larger urban areas there is growing concern regarding the degenerative diseases (heart, stroke and cancers) and 'accidents' (murder, stab wounds and alcohol-related illnesses). Some commentators draw attention to the rising trend of alcohol consumption and smoking, both of which are linked with higher rates of morbidity and mortality from degenerative diseases as well as increasing the risk of tuberculosis among men. Indeed, the total number of cases of TB rose from 88 268 in 1985 to 124 635 in 1990. Moreover, the prevalence rate of TB per one hundred thousand of popula-tion has increased 27 per cent, from 463 to 590 (SAIRR, 1993). The increasing incidence of TB may be related, in part, to the increasing number of HIV cases, since HIV suppresses the immune system and leads to the reactivation of latent TB.

The number of people infected with the human immuno-deficiency virus (HIV) is a growing problem (Cross and Whiteside, 1993). Although there were only 37 recorded cases of HIV in 1987, it is estimated that by 1990 the number of cases

Table 6.1. Notifiable Diseases in South Africa, 1985–91 (number of notifications)

	Cholera	Diphtheria	Measles	Tuberculosis
1985	700	46	15 738	53 129
1986	280	18	12 492	50 991
1987	34	28	21 120	53 627
1988	6	19	13 886	57 704
1989	3	12	18 267	68 075
1990	1	34	10 622	64 865
1991	2	12	2088*	67 056

*Notifications for 1991 are incomplete. However, part of the drop in the number of reported cases of measles may be due to the measles immunizations campaign.
Sources: SAIRR, 1990, p. 415; SAIRR 1992, p. 128; SAIRR, 1993, p. 294.

was between 119 000 and 168 000 (Figure 6.1a) (SAIRR, 1992; South Africa, 1991). The doubling period is estimated at between 8 and 10 months, and projections by the Medical Research Council suggest that by the year 2000 there could be as many as four million cases of HIV and as many as 250 000 cases of AIDS (SAIRR, 1993). The pattern of dissemination differs among the white and black populations. For whites transmission is mostly among homosexuals and drug users; thus it affects mostly males and consequently there is a low paediatric rate. However, among blacks it is largely transmitted by heterosexuals; consequently paediatric AIDS is common. Such transmission is likely to produce large national totals. Increasingly, AIDS is more common among blacks (89 per cent of the total), especially in the 30–39 year cohort, mainly among urban male and female heterosexuals (69 per cent). Trends are similar to those which have occurred in the rest of Africa: it has levelled off among the (largely white) homosexual population and has not, as yet, infected rural populations to a great extent (20 per cent). Highest rates are found in Natal, and a definite gradient declines towards the Transvaal and Cape.

The numbers that are infected with fully blown AIDS are much fewer, although the trends are the same (Figure 6.1b). However, since to date AIDS is not a notifiable disease in South Africa, it is possible that these statistics are an underestimation of the true prevalence. Among the white population it has mostly affected homosexuals: the number of reported cases rose from 26 in 1987 to a peak of 84 in 1989 and has since declined to 60 in 1991. By contrast, among the black population it has affected mostly heterosexuals and the number of cases has risen rapidly from six in 1987 to 272 in 1991. The number of cases of black infants with AIDS has increased from three in 1988 to 78 in 1991 (SAIRR, 1993, p. 288). These infants, acquiring the disease from their mothers, account for 14 per cent of the AIDS cases in South Africa and are unlikely to live beyond their fifth birthday.

By comparison, nearly 2.5m South Africans are already chronically infected with the hepatitis B virus (SAIRR, 1990). Although the mode of transmission is similar to that of HIV, that is sexual or parental, the characteristics of the 'at risk' population are very different: 88 per cent are rural blacks and a further 8 per cent urban blacks. In severe cases it can lead to liver failure and brain damage. Despite its serious nature, hepatitis B does not have the same stigma as HIV/AIDS, nor has it attracted the same attention. In many cases of hepatitis B complications are frequent and mortality rates of 10 per cent are not unheard of.

MORTALITY RATES

Life expectancy at birth ranges from 55 and 61 years, respectively, among black males and females to 68 and 76 years for white males and females (Bradshaw et al., 1992). Mortality rates also vary spatially and temporally. Although incom-

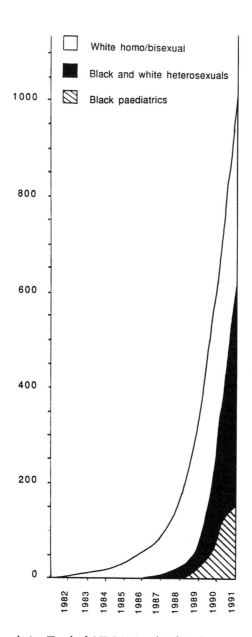

Figure 6.1. (a) Cumulative Total of AIDS in South Africa by Year and Transmission

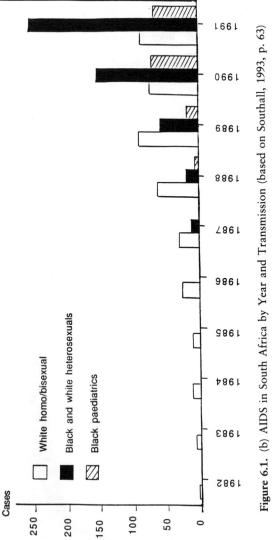

Figure 6.1. (b) AIDS in South Africa by Year and Transmission (based on Southall, 1993, p. 63)

plete, the statistics are better than those for morbidity and suggest that, for blacks, infant- and child-mortality rates are high, and the causes are mostly preventable illnesses such as diarrhoea, gastro-enteritis and respiratory infections, whereas for whites and Indians the rates are much lower and the causes of death very different. A similar differential also exists for adults.

INFANT AND CHILD MORTALITY

The infant mortality rate (IMR) is often taken as an indicator of a nation's development since it is affected by factors such as water supply, sanitation, housing, food supply and income levels. The lower the IMR, the more developed the country. In South Africa, however, reliable statistics relating to the IMR are scarce, although it is possible to see the main trends. First, it varies with race (Figure 6.2), whites having lower rates (c. 10–15 per thousand), than blacks (c. 50–100 per thousand), although the rates for both are decreasing. The latest data suggest rates per thousand of over 52 for blacks, 28 for coloureds, 13.5 for Indians and 7.3 for whites (SAIRR, 1993). The IMR also varies spatially, being higher in the periurban and rural areas compared with urban areas. Nevertheless, there is considerable variation among cities, as well as within cities, ranging from 12 per thousand in Durban to 41.3 per thousand in Port Elizabeth. Such patterns are evident in Zwelitsha, an urban area of Ciskei, where data from the Ciskei Department of Health and from the Cecilia Makiwane and Mount Coke hospitals, and Grey Hospital, King William's Town were analysed. The evidence suggested that infant- and child-mortality rates were low, less than 9 per cent,

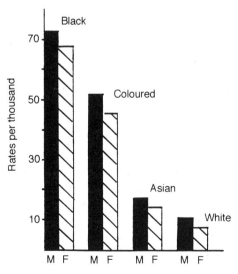

Figure 6.2. Infant Mortality Rate in South Africa, 1985

although certain areas recorded high rates, 11–17 per cent, notably in the outlying districts of Dimbaza, the former resettlement scheme and now the showpiece of Ciskei's industrial development programme.

Cause and time of death among infants also vary. For whites, neonatal (first four weeks of life) deaths were more likely, due to congenital deformities, while black deaths were more likely to be due to low birth-weight, gastro-enteritis, pneumonia and jaundice, occurring between the first and fifty-second week, the late- and post-neonatal period. At Cecilia Makiwane hospital, serving the sprawling Ciskei capital Mdantsane, the principal causes of childhood death included gastro-enteritis (20 per cent), kwashiorkor and marasmus (12 per cent) and pneumonia (12 per cent). Similar findings were found at Mount Coke and Keiskammahoek hospitals, both of which serve a more rural population, where gastro-enteritis was again the main cause of death. In each hospital undernutrition was considered to be a key condition in other deaths, notably TB and measles. In King William's Town, approximately 50 per cent of Grey Hospital's patients were in fact Ciskeian (before hospitals were 'open' to all races), and during the period 1978–88 the main causes of infant death were gastro-enteritis, malnutrition, respiratory diseases and measles. These accounted for c. 70 per cent of black infant deaths, whereas congenital malformations and asphyxia accounted for all white infant deaths. Moreover the majority of black deaths were in the neonatal period and post-neonatal period (from four weeks to one year), whilst white deaths related to the peri-natal period (first week of life) reflecting congenital malformations and immaturity. Similarly, the child mortality rates indicated high death rates from the enteritic, nutritional, respiratory and infectious diseases. Between 1976 and 1988, 85 per cent of black deaths were in these categories.

Mortality patterns among children also vary in terms of race, location and socio-economic status. For white children, mortality rates are low, consisting mainly of congenital deformities, whereas for blacks death rates are higher, especially in resettlement areas and periurban locations, comprising mostly gastro-enteritis, respiratory infections and malnutrition (Figure 6.3). Since the mid-1980s there has been an increase in the child mortality rate linked to the deepening recession.

ADULT MORTALITY

For adults, too, blacks tend to have much higher mortality rates compared with whites and Indians, especially from TB and respiratory infections (Figure 6.3). These have also increased since the mid-1980s along with degenerative diseases, such as cancers, hypertension, cardio- and cerebro-vascular diseases (South Africa, 1991). The combination of trauma, circulatory disorders, neoplasms and 'ill-defined' conditions gives an extremely high death rate for black adults, almost twice that of whites (Bradshaw et al., 1992). Indeed, Steyn et al. (1992) show that in 1988 over one-quarter of South Africans aged between 35 and 64 years of age

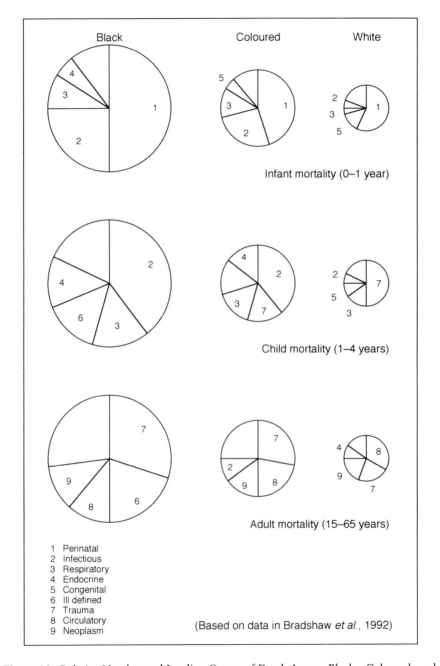

Figure 6.3. Relative Number and Leading Causes of Death Among Blacks, Coloureds and Whites in South Africa, 1985

(the potentially productive population) died of chronic diseases, that is illnesses resulting from smoking, poor diet and high cholesterol levels, which are related to lifestyle.

Thus the pattern of disease and mortality that has evolved shows that whites experience relatively more degenerative diseases, whereas blacks and coloureds are more prone to infectious, contagious diseases. However, the pattern is changing. For those in urban and periurban areas increasing incidences of stress-related diseases are emerging, and violent and 'self-destructive' causes are of growing importance. Thus the epidemiological transition model (outlined above) is a useful, albeit simplistic, method of showing how patterns of health vary and are changing in South Africa.

HEALTH SERVICES IN SOUTH AFRICA AND THE HOMELANDS

Many authors have stressed that the patterns of disease and health care relate directly to the nature of a country's development and to the unequal distribution of and access to its resources rather than to any inherent genetic differences among the less-privileged sectors of society. Just as the 'development' of disease can be related, in part, to apartheid or separatist policies, so, too, can the health care system (Mechanic, 1973; de Beer, 1984).

The history of health care in South Africa clearly reflects the nature of society, in terms of both political developments and market forces. In general, health care for whites followed (and still does) a Western-world type provision, based on curative medicine and the use of high-technology techniques. Much of this system has been privately run. Despite imminent political change, until 1994 the public health services of South Africa continued to be administered by a large number of departments, remnants of the apartheid era. These consisted of four provincial health departments, 10 health and welfare departments in the homelands and three own affairs departments for whites, coloureds and Indians. Each of the homelands had its own health budget and was able to allocate funds independently of the other departments (Figure 6.4). However, the Department of National Health and Population Development has maintained overall responsibility for the health policy of white South Africa and the so-called 'self-governing' homelands. Thus whites, coloureds and Indians were theoretically treated in segregated services, blacks in the homelands were treated by the homeland authorities and for blacks in white areas treatment was via the provincial services (which also served the other groups).

Whereas much of the health care for whites is privately funded, for some of the worst-off blacks, including those in the homelands, health care is publicly funded and provided until 1994 by the relevant homeland health service. Given their health profile a much greater orientation towards infectious diseases and pre-

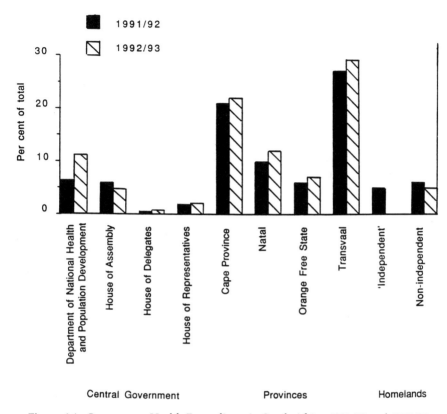

Figure 6.4. Government Health Expenditure in South Africa 1991/92 and 1992/93

ventative health care was required. Thus, the health care needs reflected the health dichotomy of the epidemiological transition: preventative, low technology for blacks compared with high-technology curative treatments for the degenerative diseases experienced by whites. However, owing to both financial constraints and the limitations of both physical and social infrastructure, health care facilities for blacks were often rudimentary and incapable of providing a comprehensive service.

Despite the reforms of the early 1990s, the health care system in South Africa still retains much of the grand apartheid plan. A number of characteristics remain: large-scale state intervention, urban bias, excessive fragmentation, white dominance and a misallocation of funds in favour of white curative hospitals and the private market (Naylor, 1988). Although hospitals became 'open' to all races in May 1990, by the end of the year only a small proportion of patients admitted to previously white hospitals were black (4 per cent).

Since the 1970s there has been an increase in state expenditure on health which can be related both to greater state participation in the economy and an increase in GNP. Although South Africa spends 5.6 per cent of its GDP on health services (compared with 7.5 per cent in the UK and 13 per cent in the USA), 20 per cent of the patients, mostly whites, consume 80 per cent of the funds (Price, 1990). Indeed, given the disease patterns of whites and blacks, and their relative population sizes, a disproportionate amount of funding is spent on whites. Expenditure on patients certainly favours whites. It is estimated that in Johannesburg R209 is spent per day on each white patient, whereas R45 per day is spent on each black patient (McIntyre and Dorrington, 1990). The shortage of funding to formerly black hospitals certainly makes their provision of care more difficult, and one of the main problems is the lack of resources which is exacerbated by the cost-spiral of medical services. Moreover, in 1990–1 the 10 homeland areas accounted for about 45 per cent of the total population but less than 20 per cent of the health budget (SAIRR, 1992).

Second, there is a huge bias towards curative medicine. Patients in the private sector receive five times as much per head as those in the public sector. Less than 6 per cent of state funds are spent on preventative medicine, and in the private sector this is even lower. Moreover, South Africa is noted for its high-technology, Western-style medical services, such as the open heart surgery at the Groote Schuur hospital, yet the majority of the population suffer from infectious rather than degenerative diseases. While curative services are certainly required, the mismatch between cure provided and cure required is evident.

Such a bias has arisen, in part, out of the market orientation of much of South Africa's health care. Consequently health care is inversely related to those in most need, that is Hart's inverse care law (1971): those who can afford it are those who least need it, whereas those who most need it cannot afford it and therefore do not get it. Thus, the provision of health care which developed was one in which there was minority control, maintained by the powerful interest groups including the medical profession, hospital industries and drug companies (Snyman Commission, 1962). Moreover, the cost of health is increasingly getting further out of the range of the poor as insurance schemes abound, medical salaries increase and the cost of technological equipment continues to rise. Indeed, the private sector experienced an increase of about 25 per cent in its costs in 1991.

A market orientation lends itself to a neglect of rural areas and homelands, since this is where there is over-representation of the poor. Many authors have commented on the urban–rural maldistribution of resources. While much of the population is rural, Benade (1992) showed that the majority of medical practitioners (77 per cent) reside in the metropolitan areas of South Africa and that the ratio of medical practitioners to population varies from 1:696 in these areas to 1:1920 in non-metropolitan areas. However, within the non-metropolitan districts there is great regional variation: in the development region of the Western Cape the ratio is 1:829, whereas in the Northern and Eastern Transvaal it rises to 1:8686. The SAIRR (1992, p. 119) provides figures of one doctor serving every 340 people in South Africa overall but only one doctor for every 15 625 in the 'non-independent homelands'.

Further inequalities between whites and blacks are numerous. For example, at least 75 per cent of whites are covered by medical aid but only about 5 per cent of Africans are, and more hospital beds and doctors are available per capita for whites than for blacks. Blacks are therefore disproportionately more dependent upon the public health services. Those in the homelands are generally worse off than other groups in South Africa, as the following figures show:

	RSA*	Ciskei	Lebowa	QwaQwa
Hospital beds/1000 population	4	4.2	2.1	2.8
Doctors/1000 population	0.6	0.4	0.0	0.1
Nurses/1000 population	4.5	3.3	1.9	1.9

(* including all 10 homelands)   (SAIRR, 1993, pp. 279–80)

The inequalities in health care are not merely a question of the number of doctors or beds per person but also concern the facilities available in hospitals and clinics, a feature which accentuates the gap. However, although it would be wrong to consider merely the quantity of resources per capita, it is impossible to assess their quality. Age, sex, qualifications, ability to speak the local language and specialist training are all factors which need to be taken into account. Thus, in a country where only 17 per cent of the population are white, over 90 per cent of the doctors are; only 12 per cent of the doctors are women and 12 per cent of doctors are over the age of 65. Less than 2.5 per cent of doctors are black and the number of black doctors graduating each year is still low (SAIRR, 1990, 1992).

A trend which has serious consequences for the development of medicine in South Africa is the continuing emigration of newly trained doctors. Recently, however, the South African medical authorities have introduced a regulation requiring all graduates to practise in South Africa for at least three years. The emigration of doctors has certainly subsided, but there are now new problems related to the influx of foreign doctors following removal of the regulation that they sit South African medical examinations prior to their entry. Following many complaints this regulation was reintroduced and nearly one-half of the doctors, mainly from Eastern Europe and South-east Asia failed to meet South African standards (SAIRR, 1993). However, there is still an exodus from the public sector, where wages are often between one-third and one-tenth of those in the private sector.

Discontent among the nursing profession is apparent. Recruitment has dropped, and there is a lack of nurses in many hospitals owing to stress, extra duties and responsibilities, low pay and long hours. Further, black nurses in formerly white hospitals have faced petty discrimination such as denial of access to the nurses' home. As a result many formerly white hospitals are operating at only 50 per cent capacity, while former black hospitals are overcrowded with patients sleeping on the floor (van Niekerk and Brown, 1989; *Weekly Mail* (Johannesburg), 10 August 1989).

Thus the provision of health care that has evolved in South Africa shows many inequalities and many constraints upon access for those most in need. An analysis of the Ciskei health services illustrates many of the comments made above and shows how the services have developed given the constraints under which they operated. The Ciskei Health Department continued to be funded by South Africa, which therefore maintained some degree of control over the practices. Moreover, South Africa also achieves a degree of control through secondment of staff, loans of equipment and admission of Ciskeians to facilities in South Africa.

The health services have a very hierarchical structure (Table 6.2). The largest regional hospital, Cecilia Makiwane, built in 1975, is located in Mdantsane. This also acts as a specialist referral hospital and as a nurse-training institution. The second tier of hospitals, including Mount Coke, developed out of small mission hospitals. Each hospital serves a number of clinics and sub-clinics in its region. These clinics represent the bottom rung of the health ladder. For example, Cecilia Makiwane hospital serves 14 urban clinics and seven rural clinics, and Mount Coke has 18 rural and four urban clinics. To these they provide doctors, ambulance services and various health personnel. The clinics are designed to offer a comprehensive health service providing general medical care, family planning, nutrition education, TB treatment, antenatal care and obstetrics. Generally, they comprise a waiting room, a maternity room, a consulting room and toilet facilities. They also offer overnight facilities; urban clinics normally have four to five beds, whereas those in rural areas generally have two. For cases which cannot

Table 6.2. Structure of Ciskei Health Services

Quaternary Level
    Referral and teaching eg Groote Schuur, Cape Town

Tertiary Level (main hospital)
    Regional hospital with comprehensive specialist and training facilities eg Cecilia Makiwane, Mdantsane

Secondary Level (minor hospital)
    Health ward hospital with limited services and/or training eg Mount Coke

Primary Level (clinic)
(a) Central clinic (high grade) eg Dimbaza, Zwelitsha
(b) Satellite clinic (low grade) eg Ndevana, Welcomewood, Peelton
(c) Sub-clinic eg Zikhova

---

be treated at the clinic patients are referred to the appropriate hospital. However, the operation of the health services is limited by a number of factors. First, the very location of clinics in Ciskei has produced a certain amount of controversy. A former Minister of Health, Mr Mzimba, stated that 'the siting of clinics takes into account population density, available infrastructure and accessibility' (Ciskei, 1989, p. 26). On the other hand, it is noted that the 'location of clinics is determined in a scientific and practical way in relation to the funds available and in terms of promises made by politicians and other office bearers' (p. 40).

In practice, these services operate under a number of constraints. Analysis by the author showed that staff shortages were a serious problem as they led to reduced efficiency and promoted a poor image of the health services. When staff levels decreased attendance at the clinics decreased, owing to increased waiting time, lack of care and attention and reduced hours of opening. Analysis of data for 1988 showed the following percentage shortfalls:

Village health workers	25	Nursing assistants	28
Medical superintendents	40	Student nurses	34
Doctors	20	Nurses	25
Ambulance officers	85	Health inspectors	30

In fact, only the director general, deputy director general and nursing administrators were covered sufficiently. The same picture was seen as far back as 1984. Quality of staff also raised some concerns. Poor staff training resulted in a failure to spot potentially ill people. Staff morale was low, owing to lack of resources, poor motivation and inadequate wages. Many of the clinics were severely limited in facilities, some without electricity or piped water. Other missing provisions included transport facilities, telephone, weighing scales, stethoscopes, road-to-health cards (giving records of development), vaccines and proper sewage systems. The size of many clinics was inadequate for the number of people served, and the lack of accommodation for nursing staff constrained the provision of a 24-hour emergency service. Transport difficulties were frequent, owing to the lack of equipment and the nature of the roads. A high turnover of incumbents in cabinet and senior civil service positions has been a great constraint upon the effective application of proposed health policies.

RECENT TRENDS

The effectiveness and appropriateness of the South African health care system is debatable (Jaros and Muller, 1986). The provision of health care has become linked to the ability to pay, and, as a result, the preventative and rehabilitative schemes have suffered. Situations have arisen, as with the building of the new Groote Schuur hospital or even the Bisho hospital in Ciskei, where the services offered do not have much relevance to the majority of the population. Although there is increasing support for a national health service, led by the medical profession (Retief, 1986), there are others such as Van der Merwe (1986, p. 728) who believe that [if] 'health is a gift bestowed on us by a higher authority . . . why should the taxpayer have to pick up the tab for those who choose to follow a slothful, self-destructive lifestyle?'

The piecemeal political reforms of the early 1980s which accelerated considerably in 1990 and 1991 were mirrored by calls for changes in the South African health services. Since May 1990, hospitals in South Africa have been 'open' to all races, a change which has occurred with remarkable ease. Indeed many, including those in King William's Town and Grahamstown, were already 'open' in practice, at the discretion of the medical superintendent. In May 1990 further rationalization and reorganization was announced. Three levels of health care provision were identified: first, a national level, in which the government formulated policy, set guidelines and norms, and was responsible for funding; second, provincial and homeland authorities were responsible for secondary hospitals, and third, local authorities were responsible for primary health care. In the 1994 constitution, health services are primarily a regional responsibility.

However, the government is still in a difficult position. With limited funds, and an almost bottomless pit of projects needing funding, including housing and education, there is a real pressure on the state to provide for those needs as well as, or ahead of, health facilities. Nevertheless, there are a number of options, including increased privatization, more primary health care (PHC) schemes for blacks, medical aid and insurance schemes and increased co-operation with non-government organizations. Higher wages could be paid to those working in the public sector, or the state could pay doctors for every treatment performed, that is fee-for-service. However, were such options implemented, it is more than likely that one would see an increase in cost, an element of discrimination and a detrimental effect on teaching and on preventative medicine (Price, 1988, 1990; SAIRR, 1990).

The weaknesses of curative medicine were noted by the South African government, yet the amount spent on preventative medicine has remained low. Moreover, given the impact of chronic illnesses on the mortality profile of South African adults, it is clear that there needs to be a wide-scale programme of health promotion and preventative medicine rather than the current orientation towards a cure-based approach; yet less than 5 per cent of the 1990–1 health budget was made available for primary health care (SAIRR, 1992, p. 127). Nevertheless, although there appears to be a growing acceptance for the need of a PHC-type approach in the delivery of health care there is a great danger that it could lead to a two-tier health care system, with PHC as the poor man's medicine and high-technology treatments for those who can afford them. However, the chances of preventative schemes improving the health status of the population are limited for a number of reasons. If anything, health care conditions in South Africa are deteriorating owing to rapid population growth and economic recession, coupled with a major shortage of funds. The whole idea of a comprehensive preventative health care team is annihilated by excessive fragmentation. Although the government favours privatization as a means of reducing the financial burden, this, too, faces problems (Price, 1986, 1988). For example, very few private companies would want to take over hospitals for low-income black patients and none without subsidy. Second, the state is committed to provide health care for those who cannot afford to pay. The provision of primary health care or community care is one means of reducing its financial burden. However, the development of a dichotomous health care system of PHC for the poor and curative health care for the well-off is certainly not in the spirit of the 'new' South Africa and would only tend to perpetuate the inequalities. Even the expansion of medical aid and medical insurance schemes will have little effect on vast numbers of South Africa's vulnerable population, although it will allow those with formal employment to have a greater participation in the provision of health care and reduce some of the government's financial burden. The state also appears to be relying on fee increases to improve facilities in the former black hospitals and is limiting the future development of public hospitals in order to encourage privatization. Yet such higher tariffs are only going to reduce access for a great majority of blacks, who already find tariffs difficult.

One further problem, and an escalating one at that, is the rapid increase in the incidence of AIDS and its effect on other diseases, notably TB and measles. Government expenditure on AIDS has risen to over R20 million, representing just under 2 per cent of the health budget. This is, of course, on top of the cost of treating other endemic diseases. Moreover, the medical, economic and social cost of AIDS will continue to rise well into the first decade or two of the next century as the HIV cases become translated into fully blown AIDS cases. With decreasing case fatality, the increased costs of prolonged treatment are clear.

CONCLUDING REMARKS

South Africa is not a microcosm of the global North–South First–Third World dichotomy, although its epidemiological pattern and its health care system may suggest as much. Rather, it is a Third World country characterized by rapid population growth and urbanization, high rates of unemployment and underemployment and a privileged (mostly white) minority. As such it resembles other developing countries such as Brazil or Mexico, although not the newly industrializing countries such as Taiwan or Singapore. It has many resources, great inequalities in wealth and, like Brazil, is moving away from an authoritarian government to a more democratic one. The transition is likely to be equally painful, if not more so, and is unlikely to be accompanied by rapid economic growth (for reasons made clear by other contributors to this volume). The probable consequences for health of low growth will be continued social and economic disintegration, leading to increases in diseases of poverty and despair (alcoholism, prostitution, drug addiction and violence).

Thus the pattern of morbidity and mortality in South Africa reflects that of a Third World country that is undergoing rapid urbanization and social transformation. For a minority of the population, who largely happen to be white, the illnesses experienced and the causes of death are similar to those of a typical developed country. However, for the majority the diseases and the deaths that occur are related to poor socio-economic conditions, poor housing, low incomes and overcrowding, to name but a few. Communicable, infectious diseases are rife. Among adults chronic diseases related to poor lifestyles account for a large part of the illnesses and deaths that are recorded. On top of this there are the problems, not unique to South Africa, of social conflict. The high level of violence and the increasing levels of drug abuse, including alcoholism, sexually transmitted disease and HIV suggest that the health care system is going to have increasing difficulties over the next few decades in serving the needs of the population. These point to a great need for preventative or promotive health care rather than curative health care, as has evolved in South Africa to date.

The demand for health care should increase with the number of women, elderly and young in a population, that is the vulnerable. It should also vary with the types of disease found and be available to those who need it most, but often this is not so. Given the population structure of many black communities, especially those in the homelands and the periurban areas, there is already a great demand for health care, which will escalate greatly in the foreseeable future, owing to population growth, social upheaval, poverty and the overrepresentation of vulnerable populations (Nagle, 1994a and b).

However, health services frequently reflect the nature of the society in which they occur, and in South Africa this is very clear. Health care in South Africa was totally locked into the apartheid system and has been used to reproduce and legitimize existing power situations in South Africa. Health care for the majority of the population is underfunded, overstretched and often inappropriate or inaccessible. Moreover, rural and periurban areas are seriously disadvantaged.

How far the health services will change as a result of the 1994 elections is debatable. The ANC in its draft policy guidelines favours the idea of a comprehensive integrated national health service, divided among national, regional and local authorities. The rhetoric clearly demands increases in primary health care, free and equal access to health services, co-operation between public and private sectors and increases in the number of black and women doctors. Likewise, statements from the Inkatha Freedom Party, the Pan Africanist Congress, the South African Communist Party and the Conservative Party call for more basic health care, appropriate to the needs of the majority although they differ in their assessment of the degree of state involvement that is desirable. It may prove difficult to escape from the legacy of previous decades as the future health services may well be heavily influenced by the location and nature of the existing facilities. With a very tight belt on government spending there is not much room for large-scale improvements.

Nevertheless the disease pattern that prevails in South Africa decrees that greater attention needs to be directed towards the infectious diseases rather than the degenerative ones. It is essential that the needs of the population are met and provided for with the best possible service. Some form of unified, non-racial, accessible and affordable health service is desperately needed. There are undoubtedly many difficulties in reorganizing the existing health structures to accommodate these changes, but in the new South Africa it is vital that the old mistakes are not perpetuated.

ACKNOWLEDGEMENTS

I would like to thank Angela and Rosie for their support and patience while I have been writing this chapter, and for being there; I dedicate this chapter to them.

REFERENCES

A number of journals carry regular articles on the subjects of health and health care in South Africa. These include *South African Medical Journal*, *Social Science and Medicine*, *Ecology of Food and Nutrition*, and *Development Southern Africa*. Wilson and Ramphele (1989) also provide a large number of references regarding surveys on health and health care issues.

Benade, M. (1992) 'Distribution of health personnel in the Republic of South Africa with special reference to medical practitioners', *South African Medical Journal*, 82 (4), 260–3.

Bradshaw, D., Dorrington, R. and Sitas, F. (1992) 'The level of mortality in South Africa in 1985—what does it tell us about health?', *South African Medical Journal*, 82 (4), 237–40.

Bryant, A.T. (1939) *A Description of Native Foodstuffs and their Preparation*, Government Printer, Pretoria.

Burroughs, E.H. (1958) *A History of Medicine in South Africa up to the End of the Nineteenth Century*, Balkema, Cape Town.

Ciskei (1989) *Annual Report: Department of Health*, Government Printer, Bisho.

Cross, S. and Whiteside, A. (eds) (1993) *Facing up to AIDS: the Socio-economic Impact in Southern Africa*, Macmillan, London.

de Beer, C. (1984) *The South African Disease: Apartheid, Health and Health Services*, Catholic Institute of International Relations (CIIR), London.

Development Bank of Southern Africa (1987) *SATBVC Statistical Abstracts, 1987*, Development Bank of Southern Africa, Pretoria.

Eyles, J. and Woods, K. (1983) *The Social Geography of Medicine and Health*, Croom

Helm, London.

Gluckmann, H. (1944) *The Provision of an Organised National Health Service for All Sections of the People of the Union of South Africa: Report of the National Health Service Committee, 1942–1944*, Government Printer, Pretoria.

Hart, J.T. (1971) 'The Inverse Care Law', *Lancet*, 1, 405–12.

Househam, K.C. and Bowie, M.D. (1988) 'Epidemiological factors in acute infectious infantile diarrhoea in Cape Town', *South African Medical Journal*, 73 (6), 346–9.

Jaros, G.G. and Muller, J. (1986) 'The cost and appropriateness of health care technology in South Africa: a plea for the development of a local medical industry', *South African Medical Journal*, 69 (10), 625–7.

Jones, K. and Moon, G. (1987) *Health, Disease and Society: a Critical Medical Geography*, Routledge & Kegan Paul, London.

Learmouth, A. (1988) *Disease Ecology*, Blackwell, Oxford.

McIntyre, D. and Dorrington, R. (1990) 'Trends in the distribution of South African health care expenditure', *South African Medical Journal*, 78 (3), 125–9.

Mechanic, D. (1973) 'Apartheid medicine', *Society*, 10 (3), 36–44.

Nagle, G.E. (1992) *Malnutrition in the Zwelitsha Area of Ciskei*, unpublished D.Phil. thesis, University of Oxford.

Nagle, G.E. (1994a) 'Challenges for the "new" South Africa', *Geographical*, 66 (5), 45–7.

Nagle, G.E. (1994b) 'Whither Ciskei: a South African homeland?', *Geography Review*, 8 (4), in press.

Naylor, C.D. (1988) 'Private medicine and the privatisation of health care in South Africa', *Social Science and Medicine*, 27, 1153–70.

Price, M. (1986) *Health Care Beyond Apartheid: Economic Issues in the Reorganisation of South Africa's Health Service*, M.Sc. Thesis, London School of Hygiene and Tropical Medicine, March 1987.

Price, M. (1988) 'The consequences of health service privatisation for equality and equity in health care in South Africa', *Social Science and Medicine*, 27 (7), 703–16.

Price, M. (1990) A comparison of prescribing patterns and consequent costs at Alexander health centre and in the private fee-for-service medical aid sector', *South African Medical Journal*, 78, 158–60.

Retief, F.P. (1986) 'A national health service for South Africa', *South African Medical Journal*, 69 (12), 728.

SAIRR (1970) *Survey of Race Relations in South Africa*, South African Institute of Race Relations, Johannesburg.

SAIRR (1990) *Race Relations Survey 1989–1990*, South African Institute of Race Relations, Johannesburg.

SAIRR (1992) *Race Relations Survey 1991–1992*, South African Institute of Race Relations, Johannesburg.

SAIRR (1993) *Race Relations Survey 1992–1993*, South African Institute of Race Relations, Johannesburg.

Seedat, Y. (1989) 'Nutritional aspects of hypertension', *South African Medical Journal*, 75 (4), 175–7.

Snyman Commission (1962) *Report of the Commission of Inquiry into the High Cost of Medical Services and Medicine*, 59/1962, Government Printer, Pretoria.

South Africa (1991) *1990 Health Trends*, Department of National Health and Population Development, Pretoria.

Southall, H. (1993) 'South African trends and projections of HIV infection' in Cross and Whiteside, *Facing Up to AIDS*, 61–86.

Steyn, K. and Fourie, J. (1990) 'Requirements of a coronary heart disease risk factor intervention programme for the coloured population of the Cape Peninsula', *South*

African Medical Journal, 78 (2), 78–81.

Steyn, K., Steyn, M., Langenhoven, M.L., Rossouw, J. and Fourie, J. (1990) 'Health actions and disease patterns related to coronary heart disease in the coloured population of the Cape Peninsula', *South African Medical Journal*, 78 (2), 73–7.

Steyn, K., Jooste, P.L., Bourne, L., Fourie, J., Badenhorst, C.J., Bourne, D.E., Langenhoven, L.E., Lombard, C.J., Truter, H., Katzenellenbogen, J., Marais M. and Oeleofse, A. (1991) 'Risk factors for coronary heart disease in the black population of the Cape Peninsula: The BRISK Study', *South African Medical Journal*, 79 (8), 480–5.

Steyn, K., Fourie, J. and Bradshaw, D. (1992) 'The impact of chronic diseases of lifestyle and major risk factors on mortality in South Africa', *South African Medical Journal*, 82, 227–31.

Tatley, M. and Yach, D. (1988) 'Diarrhoea in the Mamre community', *South African Medical Journal*, 74 (7), 339–41.

Turton, R.W. and Chalmers, B. (1990) 'Apartheid, stress and illness: the demographic context of distress reported by South African Africans', *Social Science and Medicine*, 31 (11), 1191–200.

Van der Kemp, J.L. (1804) *Transactions of the London Missionary Society, 1795–1802*, London.

Van der Kemp, J.L. (1812) *Memoir of the late Reverend J. Van der Kemp—Missionary in South Africa*, Missionary Society, London.

Van der Merwe, P.R. (1986) 'A national health service for South Africa', *South African Medical Journal*, 69 (12), 728–9.

van Niekerk, J.P. de V. and Brown, U. (1989) 'Nursing in the Republic of South Africa—a perspective', *South African Medical Journal*, 76: 525–6.

Vogelman, L. (1990) 'Psychology, mental health care and the future: is appropriate transformation in post-apartheid South Africa possible?', *Social Science and Medicine*, 31 (4), 501–5.

Wilson, F. and Ramphele, M. (1989) *Uprooting Poverty: the South African Challenge*, Norton, London.

7 Responses to crisis: Redistribution and Privatization in South African Schools

ANTHONY LEMON

Although education cannot transform the world, the world cannot be transformed without education (Robinson, 1982, p. 31)

There are no educational coups d'état (Sebidi, 1986).

INTRODUCTION

Since the Soweto revolt of 1976, and more especially since 1985 when a localized State of Emergency was first proclaimed and 'people's education' was conceived as an element in the liberation struggle, black education has been contested terrain in South African urban areas. Apartheid education has manifestly failed as an instrument of social control, a function which has yet to be restored. Today education is arguably 'the most important and intractable issue in South Africa's social fabric' (Lee *et al.*, 1991, p. 156).

The focus of this chapter is primarily on resources, which operate as a fundamental constraint on transformation of educational opportunity for the poor. But we should recognize at the outset that the education crisis, while rooted in the inequalities of apartheid, has come to reflect concerns about the content, style, structures and objectives of school education. These are reflected in the brief history of the education struggle which follows. Massive increases in the black education budget since the 1970s did nothing to satisfy their recipients, as long as the state resisted the central demand for a single education department administering a common school system for all South Africans. It later conceded the inevitability of such a development,[1] but argued that it must await the (negotiated) replacement of the 1984 constitution. This leads to the paradox of a government which in 1991 repealed the Population Registration Act and thereby deprived itself of the ability to classify people by race but which retained an educational system wholly dependent on such classification. Only in 1993 was it announced that the education system would be unified from April 1994.

The National Party government used the time it had to introduce reforms, especially in the white education sector. It will be argued that in moving towards

The Geography of Change in South Africa, edited by Anthony Lemon.
© 1995 by the Editor and Contributors. Published in 1995 by John Wiley & Sons Ltd.

an element of privatization, the government was seeking to protect the whites from the otherwise inevitable consequences of desegregation and redistribution. Such policies are likely to contribute, together with the private sector *per se*, to the gradual replacement of racial divisions with class divisions. But even within the state sector there remain major geographical constraints on the redistribution of resources which relate, *inter alia*, to demography, the location of educational infrastructure and problems of teacher redeployment. Particular attention is given to huge but relatively neglected urban–rural disparities, which it is argued should be at the forefront of future redistributive strategies. If these disparities are to be addressed, finite resources must limit what the state can afford to provide as a universal right. Beyond such provision, socio-economic differences are bound to influence educational opportunity, and strategies to minimize disadvantage will need to be considered.

INHERITED STRUCTURES, LIBERATION STRUGGLE AND STATE RESPONSES

The inequalities inherent in apartheid education are well known and the justifications of its principal architect, Dr Hendrik Verwoerd, much quoted: in essence, the 'native' was to be prepared for his station in life. In purely quantitative terms the introduction of 'Bantu Education' in 1953 led to a major expansion of black primary education, but minimal accompanying growth of secondary schools, such that by 1970 a mere 9.4 per cent of all black pupils were in secondary schools compared with 36.4 per cent of whites (Pillay, 1990, p. 34). The black dropout rate in higher forms was such that 44 white students matriculated for each black student. Most black children left school equipped only for manual labour.

During the 1950s and 1960s, black educational expenditure was tied to black taxes, under a fixed formula of R13 million plus 80 per cent of black taxes, creating a negative feedback situation. The consequences, in terms of underqualified and poorly paid teachers, overcrowded classrooms, double sessions and the costs of textbooks and stationery (free in white schools), together with high dropout and failure rates, were faithfully recorded year by year by the South African Institute of Race Relations, whose annual *Survey* remains a key source of data measuring racial inequalities in South African education.

Black pupils and teachers were also forced to use an overtly biased and racist syllabus. A new insistence by the Department of Bantu Education that mathematics and social science be taught and examined in Afrikaans was the catalyst which sparked the Soweto revolt in June 1976 and thus marked the beginning of a fierce and unresolved struggle. Widespread boycotts followed, mainly by secondary school pupils. An escalating crisis during 1977 created an atmosphere of hostility and fear in township schools.

The state responded within the framework of 'separate development'. It made cosmetic changes such as the replacement of the term 'Bantu' with 'black' in new educational legislation, and it provided more money for black education, seeking to produce more educated and trained blacks for the homelands, more trained blacks for the economy and more trained teachers to cope with rapid expansion of black education (Samuel, 1990, p. 23).

Further student protests followed in 1980, significantly relating their educational demands to the broader liberation struggle. This time the state implicitly recognized the need for major policy changes and directed the Human Sciences Research Council to investigate and make recommendations. It did so in the De Lange report (HSRC, 1981), which proposed parity of expenditure in a unitary education system but 'firmly locked within a segregationist framework' (Nasson, 1990a, p. 58). The government's belated response in a 1983 White Paper laid the basis of present education structures in terms of the 1984 constitution (South Africa, 1983). It created a new national education ministry to handle aspects of education which affect all groups, including norms and standards for financing education, salaries and conditions of staff. Otherwise, education was deemed the 'own affair' of each population group. In practice this meant the creation of separate white, coloured and Indian education ministries in the tricameral parliament but the anomalous retention of black education in non-homeland areas as a 'general affair' under its own national education department; each homeland, whether self-governing or 'independent', had its own education department (Figure 7.1).

These changes won little support from blacks. Student protest against the educational system intensified in 1984 and 1985, spreading to new regions. Students increasingly linked their grievances to wider community and political issues such as the withdrawal of the police and army from the townships, the resignation of community councillors, rent reductions and the reinstatement of dismissed workers. The Congress of South African Students (COSAS) played a key role in the mobilization and organization of support until its banning in August 1985 which led to an intensification of the boycott and the effective collapse of black education in many areas. The slogan 'Liberation now—education later' demonstrated black preparedness to sacrifice schooling in the cause of the wider struggle; it also reflected an unrealistic optimism that fundamental change was imminent.

In December 1985 the recently formed Soweto Parents Crisis Committee attempted to resolve what had now become a nationwide breakdown in black schooling by holding a national conference. From this conference the National Education Crisis Committee (NECC) emerged, and the concept of 'people's education' was born. Initially it was concerned less with the content of people's education than with the mobilization and organization of teachers, students and parents as a political force in an important sphere of the liberation struggle (Unterhalter and Wolpe, 1991, p. 11).

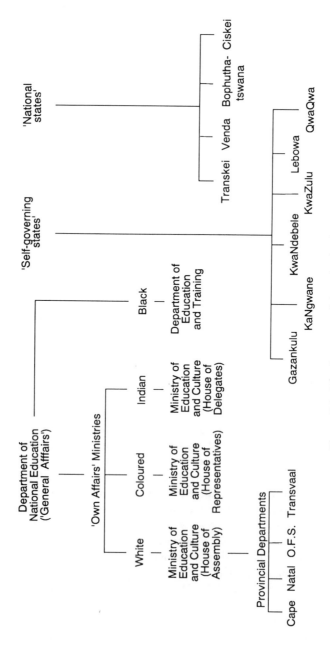

Figure 7.1. The Structure of Apartheid Education

Repressive use of the nationwide State of Emergency in the years 1986–90 succeeded in diminishing people's education as a political movement. The NECC was virtually banned, and its programme of action to assert people's power in education became impossible to implement. Academics and others worked on developing the content and character of people's education, but this effort was essentially detached from the wider struggle. The state tried to redefine the concepts of people's education, stressing the upgrading of facilities and technical issues of upgrading skills. Its changes remained within the very system which the NECC sought to displace.

In December 1989 the NECC resolved to restructure itself as a broader co-ordinating committee rather than merely a crisis committee, changing its name accordingly. At the same time it launched a back-to-school campaign. This was powerfully reinforced in February 1990 by Nelson Mandela, speaking at a large rally in Johannesburg shortly after his release from prison, when he called on all children to go back to school in order to show responsibility as future leaders and as 'the best way of welcoming me'. He said that while the quality of black education was inferior, pupils should study in spite of the difficulties they encountered (SAIRR, 1990, p. 773). His message underlined the dichotomy inherent in black attitudes to education, which is seen both as a gateway and a barrier: 'an emotional ambiguity, deep and pervasive, that has to be added to the normal clash of interests involved in policy and system change in other countries' (Lee et al., 1991, p. 151).

CONCESSION AND COMPROMISE: STATE POLICIES 1990–92

The state's educational policies have done little to assuage this ambiguity since the lifting of restrictions and the unbanning of the ANC and other groups in February 1990. Two years later education accounted for most of the laws remaining on the statute book which still embodied discrimination in terms of race. There was no real liberalization, let along the democratization demanded by the NECC and other groups (which is not without problems in the education sphere). Instead education was still very obviously in a phase of concession and compromise, giving ground slowly in response to changing perceptions of the state's own interests as well as interest group pressures (Lee et al., 1991, p. 153). Nowhere is this better reflected than in the twists and turns of state policy towards the admission policies, financing and management of white schools. As these developed during 1992, however, they began to develop a longer-term significance, both in terms of what the government was seeking to save for whites and as pointers to the socially divisive consequences of resource limitations.

In the late 1980s, rigid state school segregation led to an increasing surplus of school places for whites. Despite the closure of 203 schools between 1981 and

1991, the number of unfilled places increased from 153 637 in 1986 to 287 387 in 1991 (Metcalfe, 1991, pp. 13–14). These unfilled places resulted in part from a decline in the white birth rate but also from changing demographic patterns in inner city and some suburban areas, including the 'greying' of certain white residential areas in the closing years of the Group Areas Act. White schools in such areas were threatened with closure while parents of other race groups were forced to bus their children to schools for their own race group, even when racial mixing became legal in some of these areas under the Free Settlement Areas Act of 1988. In the areas most affected, such anomalies led to some white parental and school authority pressure to allow white state schools to admit pupils of other races.

Whereas the state's first reaction to black pressures was usually repressive, white pressures required a more positive response. This came in the introduction of three new school models in September 1990: schools could opt to go private, with a 45 per cent subsidy (model A); state-aided, with the state paying staff salaries (model C); or to manage their own admissions policy, subject to conditions intended to preserve their cultural character but remain fully state-financed (model B). Schools wishing to change status were required to hold an official poll of parents; an 80 per cent turnout was required, and 72 per cent of those entitled to vote had to support the change. Even then ministerial consent depended on a number of other considerations (South Africa, 1990). Given the sensitivity of the issue for most whites, it was widely assumed that such conditions would discourage adoption of the models in all but a handful of schools. The new policy clearly represented a deeply cautious, controlled approach to desegregation.

The results were unexpected. Just over 10 per cent of all white schools voted to change at the earliest opportunity, and a second round of ballots had increased this figure to 23 per cent by August 1991, when a further 9 per cent of schools were waiting to vote and many requests to do so were still being received. The distribution of schools opting to change was overwhelmingly urban (Figure 7.2). It was also largely confined to English-medium schools: by August 1991 no less than 95 per cent of these had voted for change in the Cape province, 79 per cent in Natal and 38 per cent in the Transvaal. By mid-1992, however, an increasing number of Afrikaans-medium schools were requesting ballots.

Official policy moved to some extent with the tide of white opinion. Ministerial permission soon became automatic when the ballot requirements were satisfied and was granted in many 'near-miss' cases too. In what amounted to a minor U-turn, a new model D was announced in August 1991 which would allow white schools with seriously decreasing enrolments to be transferred to black education departments or transformed into open schools under the control of the (white) Department of Education and Culture (DEC). By the beginning of 1993, however, there were only 17 model D schools under DEC control (Education Foundation, 1994).

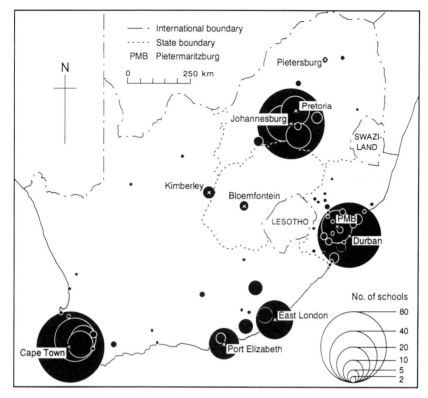

Figure 7.2. Schools Voting to Change to Model B or Model C, August 1991

How is the extent of positive responses to be interpreted? It extended well beyond schools threatened with closure, while the latter factor had not noticeably weakened Afrikaner parents' attachment to segregation. In the rapidly changing political climate prevailing since February 1990, English-speaking parents at least appear to have accepted the inevitability of desegregated schools. Some undoubtedly welcomed this, but more probably embraced change for conservative reasons, believing that control over their own admissions policies could enable more measured change than might ultimately be forced on other state schools. Certainly the admissions policies adopted suggest schools anxious to protect their academic standards and, in the words of Bryanston High School, Johannesburg, their 'cultural ecology' (*Business Day*, Johannesburg, 17 June 1991). By January 1991, when the first round of admissions to model B schools was completed, only 6059 blacks had been admitted to 'open' schools—an average of only 30 per school (Bot, 1991a, p. 4).

Schools chose overwhelmingly to vote on model B: model A was totally ignored, and fewer than 50 schools, mainly in the Transvaal, opted for model C. This clearly suggests that, whatever the perceived advantages of controlling their own admissions policies, few schools or parents wished to pay for the privilege, even in affluent suburbs.

In February 1992, however, official policy changed again, driven this time, it appeared, by budgetary constraints in a time of continuing economic recession. New staffing provision scales were introduced for the (white) DEC which implied the retrenchment of 11 000 teachers. A related budget cut of 17 per cent followed. Ostensibly to minimize teacher retrenchment, the government proposed that all DEC schools should adopt model C in August 1992, unless two-thirds of parents voted against it, in which case they could retain the status quo but with fewer teachers and reduced funding. Ownership of buildings and grounds, furniture and equipment would be transferred free of charge to model C schools. They would then become responsible for maintenance, which accounted for an esti- mated 17 per cent of total costs. Fee payment would become compulsory; initial estimates suggested that the cost would be about three times the amount of the voluntary levies made hitherto, but both the size of the latter and the predicted fee levels varied widely.

The threat of reduced teaching and maintenance provision put heavy pressure on parents to accept the change, while the two-thirds' requirement increased the difficulty of rejection, especially given the limited time available to organize a ballot and influence opinion. So in May the DEC was able to announce that 95.8 per cent of its schools had 'accepted' model C.

Within 18 months the government had moved from insistence on ballot requirements clearly intended to minimize adoption of the new models to the virtual imposition of one of them. At one level this may be viewed as evidence of its uncertainty in a period of rapid transition. But the shift to model C has more profound implications in terms of both the government's long-term aims and the probable consequences of redistribution by a future government.

Reduction of the DEC budget was consistent both with the government's proclaimed strategy of equalizing per capita education spending (see below) and the undoubted constraints of the overall economic situation. The DEC's response was condemned by the Democratic Party's education spokesman as 'a frightening disposal of family silver for a short-term budgetary expedient' (*Citizen*, Pretoria, 18 June 1992), but it was arguably much more than this. Most schools could have survived the 1992 cuts in their generous teacher:pupil ratios without disaster, but what of the future? If cuts of this magnitude emanated from a white government, what could be expected from the single education department of a majority government? Viewed in this light, the cuts were the thin end of an inevitable wedge. The only way to resist this wedge of equality was a rapid move to semi-privatization, and one which, given the previous evidence of voting on models A, B and C, would have to be effectively imposed on parents who had, for the most part, not begun to realize the resource implications of equal per capita education spending. The burden on poorer white parents was eased by a state bursary scheme, for which over 75 000 requests were received, 83 per cent of them qualifying for assistance (Education Foundation, 1994, p. 2). A racially selective scheme of this nature clearly cannot be expected to continue under the new government.

By transferring assets to the schools themselves, the government undoubtedly intended to make reversal of these changes more difficult for its successor. Small wonder that the South African Democratic Teachers Union believed that the budget was being used as a smokescreen to restructure white education before a democratic solution to the entire education crisis could be found through the process of negotiation (*Citizen*, 19 February 1992). The semi-privatization of DEC schools will effectively take them out of reach of all but middle-class blacks. Class will gradually replace race as the distinguishing feature of model C schools, as it has already in much of the private sector proper. In the post-apartheid period, it is not difficult to see a class alliance arising in defence of these schools—an alliance based, appropriately enough, on the property rights transferred to parents under the new school models.

There are implications here for the whole redistributive strategy of a post-apartheid government. The determination of whites (and increasingly of a wider middle class) to maintain high standards of education provision will not disappear with the advent of a new political dispensation. They will undoubtedly find ways of doing so, inside or outside the state system. If the compromise of model C were to disappear, the return of voluntary levies, at a much higher level than in the 1980s, seems virtually certain unless it becomes illegal. In this event, a major growth of the private sector can be envisaged.

The example of Zimbabwe is instructive here. Since 1982 parents in some government schools have used management status agreements between the Ministry of Education and parent associations to avoid the full consequences of the substantial erosion in public expenditure per pupil. But the government has not provided full legal or political support for this status (Reynolds, 1990, p. 146), and many middle-class parents, including most whites, have transferred their children to the private sector. Reynolds proposes an ingenious if somewhat complex way of overcoming this duality in what he believes to be a socially acceptable way. Management status agreements, and the associated levies on parents, would receive full official support, but the levies would be taxed for the benefit of schools in poorer areas; the deterrent effect of the tax would in turn be mitigated by incentives to schools based on their performance in collecting levies (Reynolds, 1990, pp. 148–52). Such a system would be designed to induce the middle classes to stay within the national education system, so that parents could contribute their resources, management and other skills to the benefit of all within the state system.

This critique of late-apartheid policies has taken for granted overall resource constraints. It is now time to scrutinize these more closely, first in national terms and then with reference to specific constraints.

EXISTING INEQUALITIES AND RESOURCE LIMITATIONS

South Africa's notorious racial disparities in racial per capita spending have been gradually reduced, but remain substantial (Table 7.1). The black share of the overall education budget, including the 10 homelands, doubled from 24 per cent in 1979–80 to 48 per cent in 1991–2 (SAIRR, 1992, p. 193). In April 1986 a 10-year plan to move towards parity of per capita expenditure was announced. The following year the Minister of National Education said that the government had committed itself to a real increase in educational expenditure of 4.1 per cent a year for the next 10 years, and had drafted formulas designed to bring about parity, but this would not necessarily have been fully achieved within the period of the plan (SAIRR, 1988, p. 150). Political and economic realities forced the government to announce in April 1989 that the plan could not be implemented (SAIRR, 1990, p. 788), but the ratio of white to black per capita spending did continue to decrease from 5:1 in 1985–6 to 4:1 in 1988–9 and 3.6:1 in 1991–2 (SAIRR, 1993a, p. 588).

The implications of such inequalities are revealed by a multitude of data on enrolment, teacher:pupil ratios (TPRs) and teacher qualifications. TPRs in non-homeland areas in 1989 were 38:1 for blacks, 18:1 for coloured people, 19:1 for Indians and 14:1 for whites (SAIRR, 1992, p. 205). Classroom:pupil ratios for blacks were substantially worse than TPRs. The percentage of school pupils receiving secondary education in 1991 was 27.3 for blacks and 27.5 for coloured people, but 37.5 for Indians and 41.6 for whites (SAIRR, 1993a, p. 604). Contrasts

Table 7.1. State per capita Expenditure on School Pupils by Race, 1969/70–1991/2

Year Amount	Black Amount	% of White	Coloured Amount	% of white	Indian Amount	% of white	White
1969/70	25	5	94	20	124	27	461
1979/80	91	8	234	20	390	33	1169
1983/4	234	14	569	34	1088	66	1654
1988/9	765	25	1360	44	2227	72	3082
1989/90	930	25	1983	53	2659	71	3739
1990/1	1194	29	N/A		3109	76	4103
1991/2	1248	28	2701	61	N/A		4448

Notes:
1 Capital expenditure is included
2 Figures for blacks include the ten homelands
Sources: SAIRR (1992), 195; SAIRR (1993a), 588.

in the proportion reaching standard 10 (the final year) are much greater; only 2.7 per cent of blacks and coloured people, compared with 5.7 per cent of Indians and 7.7 per cent of whites (SAIRR, 1990, pp. 824–5). Dropout rates for blacks are such that of 10 000 blacks who start school, only 113 pass matriculation (leaving) examinations.[2]

The poor qualifications of many black teachers constitute another major problem. In 1990 only 53 per cent of secondary teachers in DET schools met the official minimum requirement of standard 10 plus three years of teacher training, although this had improved from 42 per cent two years earlier; the equivalent figure for the self-governing homelands was 50 per cent (SAIRR, 1992, p. 209). In primary schools 15 per cent lacked even the former minimum qualification for black teachers of standard 6 and a diploma, a figure rising to 22 per cent in the six self-governing homelands.

Recent black matriculation rates demonstrate the effects of disruption in township schools (Table 7.2), in terms of both the low overall black pass rates and the marked geographical variations which occurred. In 1990 the disastrous results in troubled Witwatersrand schools stand out, whereas in Natal and the northern Transvaal, regions where fewer teachers participated in strikes and sit-ins, results were well above the national average, as they were in KwaZulu and the four 'independent' homelands. Regional variations outside the homelands were smaller in 1991, but Johannesburg registered only a small improvement, which was largely lost again by 1993, when the Highveld and Orange–Vaal regions did equally badly. Natal registered a further improvement in 1993, when it became the most successful non-homeland region for black pupils. Among the homelands, Bophuthatswana achieved a remarkable two-thirds' pass rate, almost maintained in 1993, while Venda and Gazankulu also showed marked improvement which they maintained in 1993.

Overall resource constraints are best captured by the fact that in 1993–4 21.4 per cent of budgetary expenditure, a figure equal to 7.3 per cent of GDP, was allocated to education. No government is likely to increase these figures significantly. Improvements in state education for blacks are thus dependent on redistribution within the global education budget, and/or on renewed economic growth. No one seriously believes that current white per capita spending levels can be extended to the whole population: 1992 estimates by Senbank and the University of Pretoria suggest that this would consume 42 per cent of budgetary expenditure and 11 per cent of GDP (SAIRR, 1992, pp. 196 and 226), while a projection by the Education Foundation puts the figure at 50 per cent of the budget by the year 2000 (SAIRR, 1993a, p. 589).

In June 1991 the government committee investigating an education renewal strategy (ERS) published a wide-ranging discussion document (South Africa, 1991). Like the De Lange report a decade earlier, it treated education as an essentially technical problem (Nasson, 1990a), paying little overt attention to political demands or the redressing of historical imbalances. It was not specific in

Table 7.2. Matriculation Results

(i) results by race group, 1988–93 (% pass rate)

Year	Black	Coloured	Indian	White
1988	57	66	95	96
1989	42	73	94	96
1990	43	79	95	96
1991	46	83	95	96
1992*	44	86	95	98
1993**	39	86	93	95

* 1992 and 1993 figures exclude results of supplementary examinations.
**African figures for 1993 refer to those candidates whose results were available; the results of about 10 per cent were still outstanding: these are excluded from all columns.
Sources: SAIRR, 1993b, p. 2; Education Foundation, 1994, p. 7.

(ii) Black results by region and homeland, 1990, 1991 and 1993 (% pass)

DET schools	1990	1991	1993	Self-governing homelands	1990	1991	1993
Cape	34	40	38	Gazankulu	36	48	50
Diamond Fields	35	42	41	KaNgwane	38	38	32
Highveld	31	37	29	KwaNdebele	29	31	25
Johannesburg	26	31	28	KwaZulu	43	38	42
Natal	41	43	48	Lebowa	28	32	29
Northern Transvaal	44	44	41	QwaQwa	31	39	31
OFS	28	42	35				
Orange–Vaal	38	40	28	*'Independent' homelands*			
				Bophuthatswana	52	66	63
				Ciskei	43	48	35
				Transkei	44	N/A	44
				Venda	40	55	53

Sources: SAIRR 1992, p. 208; SAIRR 1993a, pp. 606–7; Education Foundation, 1994, p. 7.

terms of financial provision or means of implementation (Christie, 1991). Nevertheless, most of its recommendations were at least feeling the way towards a unified, nonracial and more decentralized education system and showed awareness of resource constraints. Many of its recommendations were similar to those of De Lange, but in so far as this document was formulated by the education departments of government instead of an outside body, it did reflect a departure in official thinking (Bot, 1991b).

From a resource standpoint, the document's key proposal for school education is that state expenditure should be linked primarily to a minimum period of compulsory education, including one year's compulsory pre-primary education to help deprived children bridge the gap to school, and the current seven years of primary school; this should be extended as circumstances permit. Until 1994 there was no compulsory education for blacks, except at schools where parents requested this, whereas secondary education was compulsory for whites and coloured people (to the age of 16 or until they have passed standard 8) and Indians (to the age of 15). 'Exit points' from the formal education system are identified at standards 5 (the end of primary education), 7 and 10 (matriculation), leading to further structured education in the non-formal sector—vocational training on the job leading to certification. In the senior secondary phase (standards 8–10), subject packages could be constructed with an emphasis on either generally or vocationally orientated education.

Management councils would be established at all schools, with greater autonomy but also with responsibility for finance to supplement state provision. Building standards would be scaled down. Most importantly, in secondary schools, the assumption of the ERS document is that costs would be borne mainly by parents and the private sector.

Such a strategy would by no means guarantee equality of opportunity in practice. The financial responsibility of management councils would effectively convert all South African schools to the equivalent of model C schools and lead to wide variation in overall levels of provision according to the financial capacity of the community. The constraints responsible for such a proposal are indisputable, but it could perhaps benefit from attention to Reynolds's (1990) Zimbabwean proposals, discussed above. Those leaving the formal system at the earlier points proposed would almost certainly be overwhelmingly children of poor families, rather than those more suited to this channel. Within the secondary schools themselves, a similar danger exists that the choice between general (a euphemism for academic?) and vocational subjects would frequently be made more in terms of socio-economic background than aptitude.

The ERS discussion document was followed by publication of the ERS itself in January 1993, just one month after publication of the reports of the National Education Policy Investigation (NEPI), carried out under the auspices of the National Education Co-ordinating Committee (NECC). Both stress nonracialism and a single education ministry, with regional tiers of authority, NEPI also proposing a local tier. The ERS proposes nine years of compulsory education (two more than in the discussion document) but retreats from state responsibility for pre-school education; NEPI, like the ANC in its election manifesto, proposes 10 years compulsory education and increased subsidy of pre-school education. The ERS continues to stress vocational training, whereas NEPI emphasises a high level of general education. NEPI proposes a strong state role in eliminating adult illiteracy, whereas the ERS stresses the role of employers and community organizations.

Criticism of the ERS, NEPI reports or any other education strategy is all too easy in the context of inherited inequalities and the highly politicized nature of the education debate. All strategies must, however, include a plan for funding education fairly and with maximum effectiveness within the resources available: in the words of F.W. de Klerk himself, when Minister of National Education, 'we shall have to provide better education with fewer resources per client by means of a more efficient and leaner system of education' (*Business Day*, 15 March 1990). The final section of this chapter gives more detailed attention to some of the less publicized constraints which will act upon the adoption and implementation of an education strategy by a post-apartheid government.

GEOGRAPHICAL AND OTHER CONSTRAINTS ON REDISTRIBUTION

Demography is the greatest constraint on the potential for redistribution in South Africa's education system. Population ratios alone suggest that redistribution from the privileged sector will be thinly spread, and this holds true even if (as seems inevitable) Indians and coloured people end up on the 'losing side' of a nonracial system: the three groups combined account for only just over a quarter of the total population. Redistribution will also have to take into account those blacks of school age who are not in school: estimates vary from 1.5 to 6 million.[3] This is at the end of a decade in which black pupil enrolment increased from 4.8 million in 1980 to 8.1 million in 1991 (DBSA, 1992, p. 3): already blacks represented 80 per cent of total enrolment, and more than one black person in four was at school. It has been estimated that the number of black pupils will rise to 9.3 million by 1995 and 13.9 million, or 84 per cent of all school pupils, by the year 2000 (RIEP, 1989).

Expansion of this order will severely limit growth in per capita spending. It also poses major problems of teacher supply, especially of qualified teachers. To some extent these can be eased by a more equitable distribution of teachers within a unified education system. This implies equalization of TPRs at a much less generous level than that now existing for whites, Indians and coloured people. The World Bank (1988) argues that changes of class size within the range 25 to 50 pupils have very little effect on performance. Using this evidence, Moulder (1991a) shows that use of a TPR of 1:35 to allocate teachers in all departments in 1987 would actually produce a surplus of 20 813 teachers; a TPR of 1:40 would increase the surplus to 44 764.

A unified education department will end the absurdity of retrenching teachers in the employ of the (white) Department of Education and Culture when they are badly needed elsewhere. However, there must be serious doubt that many serving teachers will be prepared to move to black schools, or indeed to remote rural schools, as an efficient deployment of resources will certainly demand. For married women who comprise an important part of the teaching force, such moves may be impossible, while the image of township schools will prove a major deterrent to most whites. The equalization of TPRs is therefore likely to depend on moving pupils rather than teachers; this also makes sense in terms of the availability of physical facilities, but it will not be without problems. Even in schools which remain predominantly white, the doubling of class sizes implied by Moulder's TPRs would undoubtedly lead to some resignations by teachers used to a gentler professional existence. This in no way questions the logic of equalizing TPRs, but it does mean that this will not in itself wholly resolve the quantitative aspects of teacher supply, let alone the qualitative problem, which could worsen in the short term with the loss of qualified white teachers.

If people can be immobile, school plant are unavoidably so. The repeal of the Group Areas Act in 1991 was essentially a passive measure, in the sense that the apartheid cities which were so ruthlessly forged will not be eradicated for decades to come (Lemon, 1991). The best school buildings and grounds will remain where they are now, in suburbs likely to remain predominantly white, while many black secondary schools will remain without science laboratories for years to come. Equal expenditure will shift the burden of maintenance to parents in the affluent suburbs (as model C has already begun to do), but it will not lessen existing infrastructural inequality: this would require positive discrimination.

Apartheid has naturally focused attention on racial inequalities, but differences between urban and rural provision are also profound.[4] Nasson (1990a, p. 76) condemned the De Lange strategy as 'servicing the needs of metropolitan growth zones', while Buckland (1982, p. 25) commented that the rural population 'seems to feature only as a problem'. For purposes of analysis, rural areas must be divided into homelands and areas under DET control which are essentially white commercial farming areas. Blacks in the latter generally receive the worst educational provision of any group in South Africa (Graaff, 1991; Graaff and Gordon, 1992). They are dependent on farm schools which are opened (and closed) by farmers themselves, and many newspaper reports bear testimony to their precarious existence. In 1988 a mere 1.43 per cent of the half-million pupils at farm schools were receiving secondary education. Large areas of the country, such as the Border region and northern part of the eastern Cape, have no secondary facilities at all for blacks (*Daily Dispatch*, East London, 25 June 1991). The government subsidized transport to urban schools for white, coloured and Indian children living on farms but not for blacks, who must often walk long distances even to the nearest farm schools.

In 1988 the DET announced that it was increasing its contribution to physical facilities from 50 to 75 per cent; farm owners in certain border areas have received subsidies of 75–80 per cent (SAIRR, 1992, p. 202) as part of the state's policy of supporting farms in these regions for strategic reasons. In March 1993, the general subsidy level was raised to 100 per cent and extended to cover the building of toilets, fences and housing for teachers and the provision of water and electricity. The DET also paid the salaries of teachers, many of whom are unqualified, and instituted a programme of management training for all farm school teachers. The DET subsidy per pupil in farm schools belatedly began to improve in real terms but still lagged far behind even township schools. It rose from R261 in the 1988–9 financial year (barely one-third of the average in all DET schools and less than one-tenth of white per capita spending) to R647 in 1990–1 in comparison with a DET average of R1046 (SAIRR, 1990, p. 820; SAIRR, 1993a, p. 601).

The position in the homelands varies considerably. They are, of course, by no means wholly rural and include huge functionally urban areas of informal settlement within commuting distance of cities such as Durban, East London and Pretoria which are growing so fast that school provision lags far behind. The particularly rapid growth of greater Durban thus contributes to exceptionally high 1992 TPRs of 1:52 (primary) and 1:41 (secondary) in KwaZulu. The most favourable TPRs are 1:32 (primary) in Bophuthatswana and an exceptional 1:18 (secondary) in Transkei, although the latter had by far the worst primary TPR at 1:72 (SAIRR, 1993a, p. 605). In the six self-governing homelands 22 per cent of primary school teachers and 10 per cent of secondary teachers were unqualified in 1990 compared with 15 per cent and 3 per cent, respectively, in DET schools (SAIRR, 1992, p. 209).

The greatest inequalities are in terms of buildings and equipment. When salaries are deducted from per capita expenditure, little remains in many of the homelands. Science laboratories are almost non-existent. Desperate classroom shortages are reflected in 1991 primary classroom:pupil ratios of 1:70 (Transkei), 1:65 (KaNgwane) and 1:62 (Gazankulu) and secondary school ratios of 1:59 in KaNgwane and Lebowa (SAIRR, 1993a, p. 606). All these figures do, however, represent a significant improvement on the situation three years before. Other homelands were better off, especially Transkei and Bophuthatswana in terms of secondary school provision. The corresponding figures for DET schools in 1992 were 1:42 (primary) and 1:36 (secondary).

The situation in both farm and homeland schools is considerably worse than these figures suggest because they exclude children who are not at school at all (see above), the majority of whom are in rural areas. If the rural–urban differences in funding and infrastructure are to be closed, then Moulder (1991a) is probably correct in arguing that black urban education is as well funded now as it is going to be. In other words, resource constraints are likely to dictate average per capita spending in all state schools in a unified education system no greater than current levels of DET spending in the townships. Given that these townships are the very places where the educational struggle has been strongest and most politicized, and their schools the ones which have attracted international media attention to educational inequality in South Africa, this is indeed a sobering thought. Yet the need to ensure that 'rural Africans will not be ignored or short-changed when power and wealth are fundamentally redistributed in a transformed society' (Beinart and Bundy, 1987, p. ix) is as pressing as the removal of racial discrimination from the education system.

Finally, mention must be made of a less geographical aspect of inequality, albeit one where rural areas again fare worst. This concerns the very low numbers of black pupils taking mathematics and science, technical and even commercial subjects. In 1989, a mere 0.4 per cent took technical courses. In 1990, 35 per cent of black candidates sat mathematics examinations, but a mere 6 per cent passed compared with a 58 per cent white pass rate. Of 23 per cent sitting physical science, 7 per cent passed compared with 41 per cent of whites (SAIRR, 1992, p. 188). In 1991 DET sources were quoted as admitting that only one in 200 black schoolchildren matriculates with a standard of mathematics adequate for study at university and technikons (*Sunday Times*, Johannesburg, 17 March 1991). Such statistics are the context of a comment in a leaked Chamber of Mines' memorandum that 'all the efforts of the DET are immaterial to the mining industry' (*Sunday Times*, 17 March 1991). Reformist efforts designed to meet changing capitalist labour needs are only beginning to overcome the inheritance of Bantu education.

In January 1993 the state president announced that the government intended to move towards a new nonracial, regionally based education system by the end of March 1994. Funds would be allocated on a nonracial basis to the proposed regional departments. The 64 Acts regulating education would be reduced to six or seven. In practice the greater part of this extremely complex task will clearly fall to the new government which has taken office after the April 1994 elections.

CONCLUSION

Five major points are worth emphasizing in conclusion: they concern the 'late apartheid' transition, the engagement of both state and capital in the resolution of the education crisis, the need to engage all stakeholders in seeking consensus, the critical role of resources for a post-apartheid government and the limits of education itself as a redistributive process.

We have seen that recent reforms and policy proposals tend towards the replacement of racial by class divisions. The long-term difficulty of avoiding some element of this prescription is undeniable, but whereas the National Party government positively aimed for such an outcome, its successor must strive to minimize it. Late apartheid reforms, and especially attempts to reduce racial disparities in education expenditure, were welcome in themselves, but they emanated from a government in whose election blacks played no part, and they continued to be implemented in terms of the apartheid education structures decreed by the 1984 constitution. Until the establishment of an interim government and the introduction of a unified education system, pressure could only continue to be directed, as Neville Alexander (1990, pp. 166–7) puts it, at 'finding, creating and exploiting spaces within the system' and so altering 'both the dynamic and the direction of the totality that confronts us'.

Some of these spaces are being filled by a multitude of different actors outside the public sector. Collectively, the private sector has been responsible for a confusing and rapidly growing number of educational initiatives and projects: co-ordination is difficult if not impossible. While all this energy could only add up to tinkering with the apartheid system, it did give the private sector some leverage to pressure the last government to move from its own tinkering to the beginnings of a longer-term strategy, however flawed, which is reflected in the ERS discussion document.

In March 1992 a collection of political, trade-union, teachers' and educational organizations convened a National Education Conference. A key aspect of the strategy developed by the post-conference working group was to engage both the state and capital in the resolution of the education crisis, through the development of a National Education Forum (Samuel, 1993), paralleling similar bodies set up in local government, housing and other spheres in the early 1990s. Eventually, after months of negotiation, a National Education and Training Forum was established in August 1993, with a mission

> to initiate, develop and participate in a process involving education and training stakeholders in order to arrive at and establish agreements on:
> —the resolution of crises in education
> —the restructuring of the education system for a democratic South Africa
> —the formulation of policy frameworks for the long-term restructuring of the education and training systems (Education Foundation, 1994, p. 1).

It is important that the spirit of co-operation and engagement represented by such forums should develop and flourish in post-apartheid South Africa, if optimum use is to be made of all available resources. In education, there is an urgent need to develop a national strategy establishing priorities which enjoy a broad basis of consensus.

Both overall and specific resource constraints have emerged all too clearly in this chapter. Equity demands not only the equalizing of per capita expenditure but the concentration of resources for buildings and equipment on deprived areas for an indefinite period. Nothing less than transformation is required. Renewed and sustained growth in the national economy will be absolutely critical if such redistribution is to mean any real improvement for the children of the poor. Without such an improvement, not only will individuals continue to be denied opportunity, but human potential will continue to be wasted and the economy will continue to suffer critical shortages of skilled and professional/managerial labour.

Finally, it is necessary to sound a note of caution. In a review of the international literature, Nasson (1990b) has assembled abundant evidence that faith in the independent capacity of expanded schooling to redistribute income is misplaced. Even transformation of educational opportunity will not of itself bring higher status and earnings to more than a fraction of the urban poor and an even smaller fragment of the rural poor (p. 97). Education *per se* will do little to undermine the recognized inequalities of any capitalist society. Yet economic growth is essential to generate employment opportunities for those who have taken advantage of new educational opportunities. It will be the unenviable task of a post-apartheid government to judge how far it can afford to modify the distribution of wealth created by capitalism without jeopardizing the growth on which both educational and employment opportunity depend.

NOTES

1. In May 1991 the Minister of National Education pleaded with opposition groups that a single education system was 'absolutely in the pipeline' and that it was no longer necessary to fight for it (*Citizen*, Pretoria, 14 May 1991).
2. Data from the Research Institute for Education Planning, University of the Orange Free State, cited in *Productivity South Africa* 16 (4), August/September 1990.
3. Estimates include 1.5 million (James Moulder, quoted in the *Star*, 27 February 1991); 3.5–5 million ('Christian Research, Education and Information for Democracy', reported in the *Citizen*, 5 June 1991); and the Southern Africa Catholic Bishops Conference estimated the figure to be six million (*Business Day*, Johannesburg, 25 June 1991).
4. Professor James Moulder of the Department of Philosophy at the University of Natal, Pietermaritzburg has been almost a lone voice drawing attention to the rural–urban divide. See SAIRR, 1990, p. 797 and 1992, p. 194 and the *Star*, Johannesburg, 10 January 1991. Dr Neil McGurk, a prominent Roman Catholic headmaster and educationist, has also drawn attention to this issue (*Star*, 6 May 1991).

REFERENCES

Alexander, N. (1990) 'Educational strategies for a new South Africa' in B. Nasson and J. Samuel (eds), *Education: from Poverty to Liberty*, 166–80.
Beinart, W. and Bundy, C. (1987) *Hidden Struggles in Rural South Africa: Politics and Popular Movements in the Transkei and Eastern Cape, 1890–1930*, Ravan, Johannesburg.
Bot, M. (1991a) 'Open white schools—event or non-event?', *South Africa Foundation News*, 17(3), 4.
Bot, M. (1991b) *Social and Economic Update 15: Special Issue on Education Renewal*, SAIRR, Johannesburg.
Buckland, P. (1982) 'The education crisis in South Africa: restructuring the policy discourse', *Social Dynamics*, 8(1), 14–28.
Christie, P. (1991) quoted in the *Star*, Johannesburg, 3 July.
Development Bank of Southern Africa (1992) *Education in South Africa. A Regional Overview: 1991*, Centre for Information Analysis, DBSA.
Education Foundation (1994) *Edusource Data News*, 5, 1–12.
Graaff, J. (1991) 'South African farm schools: possibilities for change in a new dispensation' in E. Unterhalter, H. Wolpe and T. Botha (eds), *Education in a Future South Africa*, Heinemann, Houghton, South Africa, 221–36.
Graaff, J. and Gordon, A. (1992) 'South African farm schools: children in the shadow' in R. McGregor and A. McGregor (eds), *McGregor's Education Alternatives*, Juta, Kenwyn, South Africa, 207–37.
HSRC (Human Sciences Research Council) (1981) *Education Provision in the RSA* (De Lange report), Government Printer, Pretoria.

Lemon, A. (ed.) (1991) *Homes Apart: South Africa's Segregated Cities*, Paul Chapman, London.

Lee, R., Schlemmer, L., Stack, L., Van Antwerpen, J. and Van Dyk, H. (1991) 'Policy change and the social fabric' in R. Lee and L. Schlemmer (eds), *Transition to Democracy: Policy Perspectives*, Oxford University Press, Cape Town, 127–58.

Metcalfe, M. (1991) *Desegregating Education in South Africa: White School Enrolments in Johannesburg, 1985–1991: Update and Policy Analysis*, Research Report 2, Education Policy Unit, University of the Witwatersrand, Johannesburg.

Moulder, J. (1991a) 'Unequal inequalities: teacher:pupil ratios', *Indicator SA* 8(2), 76–8.

Moulder, J. (1991b) *Facing the Education Crisis: a Practical Approach*, Heinemann, Houghton, South Africa.

Nasson, B. (1990a) 'Redefining inequality: education reform and the state in contemporary South Africa', in B. Nasson and J. Samuel (eds), *Education: from Poverty to Liberty*, 48–78.

Nasson, B. (1990b) 'Education and poverty', in B. Nasson and J. Samuel (eds), *Education: from Poverty to Liberty*, 88–108.

Nasson, B. and Samuel, J. (eds) (1990) *Education: from Poverty to Liberty*, David Philip, Cape Town and Johannesburg.

Pillay, P.N. (1990) 'The development and underdevelopment of education in South Africa', in B. Nasson and J. Samuel (eds), *Education: from Poverty to Liberty*, 30–47.

Reynolds, N. (1990) 'Planning for education expansion in Zimbabwe', in B. Nasson and J. Samuel, *Education: from Poverty to Liberty*, 141–56.

RIEP (Research Institute for Education Planning, University of the Orange Free State) (1989) *Education and Manpower Development*, 10.

Robinson, P. (1982) 'Where stands an educational policy towards the poor?, *Educational Review*, 34(1), 27–33.

SAIRR (South African Institute of Race Relations) (1988) *Race Relations Survey 1987/8*, SAIRR, Johannesburg.

SAIRR (1990) *Race Relations Survey 1989/90*, SAIRR, Johannesburg.

SAIRR (1992) *Race Relations Survey 1991/92*, SAIRR, Johannesburg.

SAIRR (1993a) *Race Relations Survey 1992/93*, SAIRR, Johannesburg.

SAIRR (1993b) *Fast Facts*, 8, 2.

Samuel, J. (1990) 'The state of education in South Africa' in B. Nasson, and J. Samuel (eds), *Education: from Poverty to Liberty*, 17–29.

Samuel, J. (1993) 'A proposition for a national strategic framework for education in South Africa', *Development Southern Africa*, 10(2), 249–61.

Sebidi, L.M. (1986) 'The current education crisis' (editorial), *Funda Forum*, May.

South Africa (1983) *White Paper on the Provision of Education in the RSA*, Government Printer, Pretoria.

South Africa (1990) *Additional Models for the Provision of Schooling: Information Document*, Department of Education and Culture, Administration: House of Assembly, Cape Town.

South Africa (1991) *Education Renewal Strategy, Discussion Document*, Committee of Heads of Education Departments, Department of National Education, Pretoria.

Unterhalter, E. and Wolpe, H. (1991) 'Reproduction, reform and transformation: the analysis of education in South Africa' in E. Unterhalter, H. Wolpe, T. Botha, S. Badat, T. Dlamini and B. Khotseng (eds), *Apartheid Education and Popular Struggles in South Africa*, Ravan, Johannesburg, 1–17.

World Bank (1988) *Education in Sub-Saharan Africa: Policies for Adjustment, Revitalization and Expansion*, World Bank, Washington DC.

8 Housing and Urban Reconstruction in South Africa

MALCOLM LUPTON AND STUART MURPHY

INTRODUCTION

In South Africa, racial segregation effectively meant that in terms of the 1910 Union Constitution as well as the Land Act of 1913 and its subsequent amendment of 1936, 87 per cent of the national territory of South Africa was allocated to whites, while the remaining 13 per cent was reserved for blacks. Moreover, as initially expounded by Lord Milner, and further elaborated by Colonel Stallard, the principal race theoreticians of the day, South Africa, unlike the colonies in tropical Africa and Asia, would evolve into a 'white man's land' (Denoon, 1973). That is to say, alongside policies directed at securing an effective white majority by means of Milner's 'demographic revolution' (Denoon, 1973) blacks were deemed 'temporary sojourners ministering to the labour needs of whites': immigrants who would return to their native lands.

While retaining classical segregationist premises regarding the spatial separation of the races and reliance on cheap black labour, post-1948 National Party governments introduced several innovations which became articulated into a coherent and comprehensive geo-political vision of apartheid. Nevertheless, apartheid as a geo-political concept faded from the moment of its inception as the day-to-day running of the South African state and economy increasingly relied on cheap black labour, causing in turn major contradictions as far as urban and housing policies for Africans were concerned. As accommodation for blacks was intimately tied to the wider geo-political system of apartheid, analysis of South Africa's shifting policies and politics of shelter provides a lens for tracing the geography of change in a racially fractured society.

First, a brief overview of the changing relationship between housing provision for blacks in South Africa and the wider geo-political concerns of apartheid urban planners is provided. In part two the intention is to examine what may be characterized as competing contemporary housing policies and urban strategies within South Africa's establishment and alternative institutions. Finally some of the key elements which may form part of a future national housing policy are highlighted in the concluding section.

The Geography of Change in South Africa, edited by Anthony Lemon.
© 1995 by the Editor and Contributors. Published in 1995 by John Wiley & Sons Ltd.

THE CONTRADICTIONS OF BLACK HOUSING PROVISION DURING THE APARTHEID PERIOD: A BRIEF OVERVIEW

Within the ruling National Party two broad schools of thought could be discerned regarding the goals and practical implementation of apartheid policies (Posel, 1992). One school, as represented by the German-educated intellectuals Drs Verwoerd and Eiselen, contended that to make apartheid 'work', whites would have to make substantial material sacrifices by accepting a general reduction in living standards, especially as far as the white working class was concerned: hence the slogan 'poor but white'. White minority supremacy over 87 per cent of the national space, and the attendant economic advantages, had to be abandoned in favour of a geographically reduced South Africa with a white, albeit poorer, majority.

For Verwoerd, and especially Eiselen, apartheid was but a temporary necessity, a way station toward a white majority state, based on white labour (Posel, 1992). As Dr Verwoerd gazed into the future he discerned, on the southern tip of a hostile black continent, the outlines of a geographically reduced Republic of South Africa resembling the great white states of Australasia, North America and Western Europe. The central paradox of this school of thought was the contention that when apartheid has achieved its objective, in the not-too distant future, the need for racially discriminatory measures, for which South Africa has earned global notoriety, would diminish and finally disappear. It was envisaged that when the National Party had completed its historic mission, a spatially smaller, but demographically reconstituted South Africa would, like the other white dominions, proudly assume its place on the stage of nations as a liberal democracy with a white majority. According to Verwoerd, apartheid, or the geographic separation of South Africa's races, was to be pursued, as a temporary but necessary expedient, with an unprecedented vigour and consistency so as to make racial discrimination redundant: 'in pursuing [racial] segregation far enough . . . [racial] discrimination can only be cancelled out' (*Hansard*, 14 April 1961, col. 4708).

Another school of thought regarded apartheid as an end in itself: securing, for as long as was possible, white minority political supremacy as well as material privilege within an economy based essentially on cheap black labour. Given the rather incompatible demands of the white electorate for racial separation together with material privilege, the second school finally prevailed. While simultaneously asserting the impermanence of the urban African in 'white' South Africa, the benefits of cheap black labour for the South African economy were emphasized, compelling even Dr Verwoerd's government to concede that in the interests of industry and general economic growth, residential stabilization of a segment of the black workforce was imperative (Hindson, 1987). Hence the appearance of a black urban population 'segmented' into urban 'insiders' and 'outsiders' (Hindson, 1987). Housing policy also emerged as a crucial

Figure 8.1. Formal Township Houses, Soweto

mechanism for segmenting the urban African workforce. From the 1950s accommodation for African urban 'insiders' was effectively provided by government as well as the private sector. Generally, housing for urban blacks was provided in two ways.

First, government was indirectly involved in housing provision by advancing finance to enable individuals and households to accommodate themselves. Although the then Department of Community Development introduced the government's new, market-oriented housing policy for all races in 1983 (Mabin and Parnell, 1983), black home-ownership in South Africa has a long and chequered history. Prior to 1955, blacks could own property under freehold title, after which 30-year leasehold was introduced (SAIRR, 1980). In 1968 however, 30-year leasehold was abolished when the Minister of Bantu Administration issued a directive to local authorities stating that blacks in 'prescribed' (that is white) areas should not be allowed to build their houses on 30-year leasehold plots or be allowed to purchase homes from local authorities. After 1968 government policies also required that blacks be accommodated within nearby 'homelands' instead of 'white' South Africa, giving rise to places like Mdantsane, for example. The ability of municipalities to build houses for blacks within their own boundaries was therefore limited. Furthermore, this policy change in the late-1960s has been a major contributor to the scale of the current shortfall in

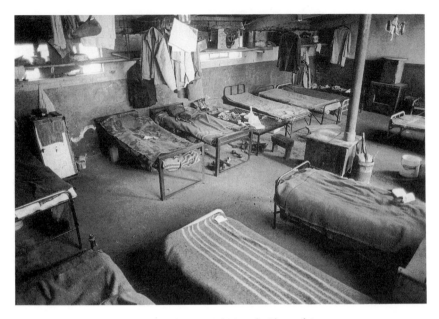

Figure 8.2. Interior of Hostel, Alexandria

housing: in 1993 the housing backlog was estimated to be as high as three million units, whereas only 10 per cent of blacks could afford to contribute to their accommodation needs.

Henceforth, blacks were allowed only to rent accommodation, a restriction removed only in May 1975 (Horrell, 1978). Although blacks were allowed, once more, to purchase houses in terms of 30-year leasehold, existing legislation at the time prevented financial institutions from granting loans for black home-ownership schemes, as the African was considered, according to prevailing apartheid ideology, a 'temporary sojourner' in the cities of 'white' South Africa. Blacks could, therefore, not obtain freehold title to sites within the jurisdiction of 'white' South Africa. Only when 99-year leasehold was introduced by the Black (Urban Areas) Amendment Act of 1978, was the Financial Institutions Act amended, thus allowing building societies and banks to advance, for the first time, housing loans to black leaseholders (SAIRR, 1980).

Second, there was direct state intervention in the production and management of the housing stock for Africans. From the 1950s, central as well as local government were instrumental in providing large numbers of four-roomed, 'matchbox' houses for renting. The late 1970s however, saw the start of a major shift in urban and housing policies toward the privatization of state-owned housing stock and a concomitant promotion of home-ownership schemes (cf.

Figure 8.3. Detached Matchbox Houses

Figure 8.4. Single-unit Matchbox House, Soweto

Bond, 1990; Lupton, 1993; Swilling, 1990). Specifically since 1983, the state's direct participation in housing provision decreased, with a greater emphasis on the role of the private sector (Mabin and Parnell, 1983). From the late-1980s market-oriented approaches to housing provision gained widespread acceptance among government, mainstream institutions and even non-governmental development agencies, to which attention now turns.

COMPETING URBAN AND HOUSING POLICY AGENDAS

CONTEMPORARY MAINSTREAM AGENCIES

Local establishment agencies do share a broad philosophy with the World Bank (Tomlinson, 1992), whose suggestions on urban housing issues will therefore be examined before exploring the policies and practices of local establishment institutions operative within the housing field.

The World Bank

The World Bank embarked on a series of information-gathering missions to South Africa, having anticipated multi-racial elections for a government of transition. The World Bank also anticipated playing a major role after the election, and is anxious to influence urban and housing policy agendas. The Bank identifies apartheid policies, such as the denial of property rights to blacks and the Group Areas Act, as mainly responsible for the housing crisis in South Africa in that such policies are deemed to have distorted the operation of rational market forces. Moreover, the Bank suggests that local policy planners should bear in mind the linkages between the housing sector and broader social and economic policies (World Bank, 1992). If the problems of growth and equity are to be successfully addressed at the national level, the ultimate objectives of housing policy reform in South Africa, the Bank claims, are not only to 'lead the way . . . but it must be managed carefully during the period of transition to a more egalitarian society' (p. 2). Thus, for the economy to function well, cities must function well, and for cities to function well, the housing sector must function well (World Bank, 1992). The Bank accordingly rejects policies which support massive government provision of completed houses of a high standard, preferring instead 'enabling strategies' whereby government facilitates the efforts of the private sector and community-based organizations to deliver housing. Within this scenario an aspect of government intervention would be capital subsidies to the most disadvantaged of society. Governments should ultimately intervene to assist market mechanisms working in the interests of all citizens (World Bank, 1992).

An essential element of housing policy on the demand side of the equation, then, should be to increase black people's access to formal financial institutions. The Bank points out that nearly 40 per cent of consolidated assets of South Africa's banking system are as residential mortgages, yet only 12–15 per cent goes to borrowers in black residential areas. Another aspect identified by the Bank is the lack of security of tenure. Programmes to upgrade and clarify tenure of informal settlements are seen as important in enabling people to mobilize resources to consolidate and upgrade their dwellings and neighbourhood facilities. Also, the Bank advocates the formulation of government subsidies which should be equitable, well targeted down market, non-distorting, measurable, transparent and simple to administer (World Bank, 1992).

On the supply side of the housing delivery equation, the Bank argues that flexibility is essential and should be informed by ensuring an adequate delivery infrastructure to ensure that households are willing to invest in housing; are rationalizing the legal and regulatory framework; and ensuring that the building industry is competitive and responsive to demand (World Bank, 1992). The Bank further mentions that the existing institutional framework for the formulation and implementation of housing policy in South Africa does not serve the interests of individuals or the economy. To this end a single housing ministry charged with overall responsibility for policy formulation, co-ordination, and monitoring the performance of the housing sector is required. The Bank envisages that this ministry should provide both an institutional linkage between housing and macroeconomic planning and a platform for participation of the various stakeholders in housing policy formulation (World Bank, 1992).

The Urban Foundation

Apart from the World Bank, the Urban Foundation (UF) is actively involved in influencing official housing and urban policy and in the delivery of low-cost housing. Since its establishment as a private-sector lobby group on urbanization in 1977 after the 1976 Soweto riots, the UF was at the forefront in influencing the government to introduce the 99-year leasehold and home-ownership schemes for Africans and in legitimizing the concept of self-help housing in the South African context. Before examining the performance of the UF's record in low-cost housing delivery, its policy proposals for addressing South Africa's housing crisis need to be sketched.

The UF, like the World Bank, favours a market-driven approach with strategic government intervention to resolve South Africa's housing crisis. Ultimately, the UF argues, the government must take sole responsibility for achieving a sustainable housing policy which allows all people to secure housing within a safe and healthy environment. The government should therefore intervene in the housing market to secure the legal and institutional conditions to enable maximum private-sector involvement in housing delivery. The UF further contends that when the public sector provides housing it must operate on a full

Figure 8.5. Nancefield Hostel, Soweto

Figure 8.6. Informal Settlement, Soweto

Figure 8.7. Free-standing Shacks, Mshenguville

cost-recovery basis to ensure that the state does not operate under conditions that are more favourable than those applying to the private sector.

For the UF, government must take responsibility for policy and establish a regulatory framework for housing provision. Such a policy would focus on the urban poor and promote, in particular, the informal housing market (Urban Foundation, 1990). The UF thus emphasizes the importance of informal housing, which it asserts can deliver at the required rate and scale. Regarding urban land for low-income and informal housing, the UF claims that the government must intervene in the land market since space for the urban poor will not be assembled by market forces alone. In this regard, the UF suggests land investment trusts be established to finance land acquisition and servicing activities of all private and public-housing agencies (Urban Foundation, 1990).

In line with World Bank thinking, the UF argues that government subsidy schemes should target the poor (rather than the middle class as happens with existing first-time home-owner subsidy, which is a subsidy on interest payments); mobilize the private sector to a maximum; be sustainable; and be simple to administer (Urban Foundation, 1990). The UF thus proposes a capital subsidy of a fixed amount on the selling price of a serviced site which would subsidize site-and-service and informal settlement upgrading schemes.

Furthermore, it is recommended by the UF that the existing first-time home-owner subsidy scheme be restructured and oriented further down market and that all other subsidies be phased out. The UF notes that private-sector financial institutions service some 15 per cent of black families in South Africa and claims that innovative mechanisms need to be formulated regarding both lending terms and in assessing eligibility for loans to provide finance for a wider market than at present. Moreover, loan finance in the informal housing market is required for consolidation purposes. Pension and provident funds should also be encouraged, the UF proposed, to explore innovative mechanisms for funding housing, particularly for those at the lower end of the housing market (Urban Foundation, 1990).

In terms of housing delivery, the Home Loan Guarantee Fund was jointly initiated by the UF and the Mortgage Lenders Association in mid-1990 and aims to encourage financial institutions down market by facilitating the provision of home loans of between R12 000 and R35 000 for conventional and starter housing. The purpose of the fund was to address two constraints on lending in the low-income market: first, the perceived risk of not recovering funds loaned; second, the non-profitability of small loans. The scheme also had as an objective the extension of housing finance to an additional 30 per cent of the population by lowering the normal lending threshold and the deposit required on a house.

Financial institutions traditionally grant loans of up to 95 per cent of the property value, and borrowers generally have to provide a deposit of 10–20 per cent of the purchase price. With the loan guarantee system however, the borrower need only supply a 5 per cent deposit, either in cash or equivalent collateral from the borrower's employer. The borrower is then required to purchase a deposit replacement guarantee (to ensure receipt by the lender of a sum equal to the full deposit of 20 per cent) and a renewable risk reduction guarantee (to cover any losses that might be incurred by the lender in execution of the sale of the property) at the same premium rate.

All these costs, including bank service charges, are amortized over the period of the loan. The loan guarantees are payable to the financial institution concerned if there is a loss in execution of the sale of the property arising from a default in repayments by the borrowers (Pretorius, 1990). The fund had insured a total of 10 236 loans by March 1993. The value of the loans granted stood at R304 million against a property value of R345 million (Pretorius, 1990). The home loan guarantee initiative was originally structured to aid the provision of R1 billion in conventional mortgage finance (approximately 25 000 individual home loans) from financial institutions.

Most of the country's major financial institutions had agreed in principle to participate in the initiative, but except for the Perm, they are not contributing significantly to financing housing below R43 000. Financial institutions contend that the main factors constraining the supply of funds for low-income housing are the perceived instability of the political environment, the negative investment mood in the country, the administrative costs of handling small loans and a perception that risks of default are higher for smaller loans than larger ones (cf. Oelofse and van Gass, 1992; Urban Foundation, 1990; Urban Foundation, 1992a).

The Independent Development Trust

In March 1990 State President F.W. de Klerk announced that a R2 billion trust fund (later named the Independent Development Trust) was to be established for socio-economic development. Mr de Klerk invited a former chairman of the Urban Foundation, Jan Steyn, to head the trust. Steyn accepted with the proviso that government promise to repeal the remaining legislation which discriminated on racial grounds and the trust remained independent of the government.

The areas which were targeted for intervention included housing, education and training, health and welfare, job creation and small business development. An amount of R750 million, subsequently increased to R845 million, was allocated for a once-off capital subsidy scheme. Some 112 690 serviced sites for approximately 800 000 people were expected to benefit from the scheme (SAIRR, 1990, 1992, 1994). The capital subsidy scheme involves the payment of a once-off capital subsidy of R7500 for each site to be developed for low-income households. The scheme applies to sites sold for the first time only and can be part of an informal settlement upgrade project or a new site-and-service scheme.

The capital subsidy is granted to the developer once the transfer of a particular site to the purchaser is registered. Only first-time homeowners over the age of 21 years with dependents living with them, and who have a monthly income of less than R1000, are permitted to qualify for the subsidy. Furthermore, a participant must receive no other housing subsidy from the government. The scheme is implemented on a project basis through commercial developers, utility companies, community-based housing associations, non-governmental organizations, local authorities and provincial administrations. Community participation is regarded as a key component of projects.

By 1993, after observing a severe lack of experience and training to manage development because of lack of opportunity, the IDT addressed the issue of community participation and local institutional capacity in a more structured manner by setting up a consolidation programme. The programme entailed the training of people as consolidation officers from communities to receive serviced sites (Cullinan, 1993a). To qualify for the subsidy, projects must also include water-borne sewerage on each site, metered water, bitumen-surfaced main roads, *in situ* formation-graded roads for minor roads and unlined stormwater drainage. The ownership of a site is regarded by the IDT as the first phase in the consolidation process, involving incremental upgrading of homes, the provision of facilities, the upgrading of infrastructural services and the initiation of socio-economic development programmes.

By 1993, and with the benefit of hindsight, the IDT acknowledged that it had been too prescriptive regarding level of service and site size. The IDT however, remains adamant that site beneficiaries must become individual owners, thus regarding alternative tenure forms as non-negotiable (Cullinan, 1993a). By mid-1993 an amount of R334 million was paid out to 92 projects which were either in the process of being carried out or had been completed. The IDT anticipated that the balance of the R511 million would be paid out by mid-1994.

Most IDT projects are located in the Pretoria/Witwatersrand/Vereeniging metropolitan complex, Natal and the Eastern Cape (SAIRR, 1994). While falling within a market-oriented approach, the IDT's once-off capital subsidy scheme has been criticized, among others, by local establishment institutions such as the Development Bank of Southern Africa (DBSA) and the South African Housing Trust.

The Development Bank of Southern Africa

The DBSA argued that the scheme created 'islands of privilege' for certain developers, in that owners of unsubsidized land had difficulty selling sites adjacent to subsidized sites. The limited choice as to the location of a serviced site, usually on the urban periphery, has also been criticized, as has the cut-off point preventing those earning more than R1000 a month from obtaining a subsidy (*Financial Mail* (Johannesburg), 22 November 1991). While commending the IDT's capital subsidy programme as a far more appropriate approach to housing subsidies than that found in most developing countries, the World Bank recommended that less capital be granted to qualifying individuals, thereby reaching a broader spectrum of the poor. In other words, 'fewer for more rather than more for fewer'. The DBSA also suggested that the subsidy be granted to the user rather than the developer. Following the World Bank, the DBSA contends it would reduce potential corruption while simultaneously stimulating competition between developers (*Sunday Tribune* (Durban), 18 August 1991).

Of the private sector actors, commercial housing developers have been the most critical of the capital subsidy scheme, and of site-and-service projects in particular. Developers maintain that the community participation requirement has excluded their participation from such projects, while benefiting development agencies with previous experience of working with community-based organizations (such as utility companies). Moreover, commercial developers claim that profit margins on the sale of serviced sites are very low while the risks on holding costs of land are high (Walker and Merrifield, 1993).

From a 'user' perspective, the IDT's capital subsidy scheme has also been criticized on several grounds. While the IDT has been able to deliver serviced sites relatively quickly and at an affordable level, the product is perceived to be inadequate and incomplete. Beneficiaries of serviced sites have also complained about the unfair demands which consumers of conventional housing do not bear: user's spare time, difficulties of living in incomplete houses and the length of time required to complete a house (Walker and Merrifield, 1993). Community-based organizations have also pointed out that the IDT granted subsidies to contractors who provided poor workmanship without consulting the community (*The Star* (Johannesburg), 6 May 1992).

PLANACT, a development consultancy to community-based organizations notes that the IDT assumes all beneficiaries will choose the same product in designing a standard product, yet simultaneously insists on community partici-pation (Cullinan, 1993b). Resources are not set aside for building beneficiary capacity to participate in technical negotiations, yet the participation of bene-ficiaries is stipulated by the IDT as a prerequisite. Furthermore, because the IDT allocates the subsidy without interest, site beneficiaries are penalized for the time taken to plan the project, since with each month that passes, inflation reduces the R7500 subsidy by some R100.

The delivery of serviced sites has also led to the perpetuation of mass urban sprawl, owing in large part to the fact that the subsidy covers the cost of land which is cheaper on the city periphery. In addition, users are often unable to afford monthly service charges which opened the possibility for downraiding (better-off residents purchase sites from poorer residents) (Cullinan, 1993b). The IDT has further established several financing mechanisms directed at providing loan finance to the poor for sheltering the market segment regarded by main-stream financial institutions as too risky and providing an insufficient return.

The South African Housing Trust

The South African Housing Trust (SAHT), a non-profit organization, was established by the government in November 1986. The initial board of trustees was nominated by the government and received a grant of R400 million (SAIRR, 1987). The aim of the SAHT is to promote the provision of affordable housing and shelter and security of tenure to the lower-income groups of South Africa through the development of serviced land and the provision of loans. In the preamble of its mission statement, the SAHT maintains that the private sector should be involved to a maximum in housing provision and aim to involve beneficiaries 'wherever possible and to the extent practical' in decision-making processes and housing delivery (South African Housing Trust, 1992). Black people were to be the main beneficiaries of the SAHT. It also intended to promote job opportunities by enabling the unemployed to build their own houses under a self-building scheme.

Since its inception, the SAHT has been involved in the delivery of approxi-mately 38 500 houses, all via private sector contractors, and the servicing of 48 000 stands were serviced by the SAHT. Through the retail lending operations of Khayalethu Home Loans some 6500 home loans with an average value of R29 000 per loan were granted during 1992, bringing the total number of clients in the scheme to 26 000 (South African Housing Trust, 1992). The Trust has tended to focus more on delivery or the 'product' aspect of the 'product/process' equation: the Trust boasts that it was the biggest contributor in the category of low-cost houses during 1992, being responsible for providing 63 per cent (9500 homes) of the total of 15 000 starter/incremental homes (35–50 square metres)

(*Housing in Southern Africa*, January 1993). However, recent statements from the trust's management appear to recognize the need to become more conscious of 'process': giving project beneficiaries more control over the development process (*Financial Mail*, 27 November 1992).

The De Loor proposals of the South African government

Finally, within the broadly defined establishment camp, the proposals of the De Loor Task Group, which was appointed by the South African Housing Advisory Council to provide recommendations to the South African government on the formulation of a new national housing policy and strategy, are examined, together with certain criticisms levelled at the recommendations. The De Loor Task Group presented its proposals to the government in May 1992. The then minister of national housing and public works, Leon Wessels, said that the recommendations would not result in legislation as the government sought to discuss the proposals with other stakeholders in the housing field (*The Star*, 27 January 1992). The government also indicated that the De Loor proposals should be seen as its contribution to discussions at the National Housing Forum (NHF) (Department of Local Government and National Housing, 1992, pp. 10–11).

Unlike public pronouncements of certain government ministers during apartheid's heyday, the South African government recently became concerned to be perceived to be following internationally recognized trends in policy formulation. Concomitantly, it has become especially responsive to policy recommendations of powerful international financial agencies, particularly the International Monetary Fund and the World Bank. Indeed, the De Loor task team has taken its cue from the World Bank's approach to housing which it describes as being in line with current international thinking on housing. The De Loor proposals essentially fall within a market-oriented framework, where housing is seen as 'an economic good' or a commodity.

Key policy principles contained in the report include the following: housing policy should be applicable to, and equitable toward, all legal residents without discrimination based on race or gender; community involvement and participation should be a part of housing policy; housing should be a part of broad socioeconomic policy and a part of urbanization, regional development and national development policies and strategies; and an adherence to market principles in housing provision, not in a *laissez-faire* manner but where the public and private sectors follow a synergistic relationship. In line with World Bank recommendations the report suggests that the government should adopt an enabling strategy, involving policy formulation, co-ordination, the creation of a regulatory framework, and the provision of key interventions, for example a once-off capital subsidy.

Where the government is already involved in housing schemes, it is recommended that market-related approaches should be adopted. The report describes the role of the private sector as having a 'superior ability' to deliver and maintain sites and housing. It is maintained that there is space for the housing sector to increase its share in the economy and the government budget, from the present share of less than 3 per cent of GDP to approximately 5 per cent.

Furthermore, it is suggested that the amount set aside for housing and related infrastructure in the central government's budget should be increased from the current average of R1.6 billion to some R3.5 billion (at 1991 prices) (South Africa, 1992). The report states that government financial support should be targeted at lower-income groups as a once-off capital subsidy, which would replace existing subsidy schemes that essentially benefit the middle-class and not the very poor. A four-phased subsidy scheme is suggested, graduated according to income, starting with the provision of a serviced site and progressing to a formal house.

The first subsidy category caters for very poor households with monthly incomes below R1000. The beneficiary receives a serviced site and pays monthly service charges yet remains a tenant of the local or regional authority, although provision is made for freehold tenure at a later stage. Category two is aimed at households earning a monthly income of below R1500. Qualifying households receive a site with basic services yet own the site once a loan of up to R5000 is paid over a five-year period. Beneficiaries also receive technical assistance in consolidation: the incremental upgrading of informal settlements. The third subsidy category is targeted at households with a monthly income of between R1001 and R2000. Beneficiaries receive a capital subsidy of R7000 and have the option of renting or owning a house, flat, site or room. The fourth category includes a capital subsidy of R6000 (which decreases as incomes increase) to households earning between R2001 and R3000 a month. Beneficiaries are given the option of renting or owning a site, house, flat or room.

To qualify for categories two, three and four, the head of a household must be a first-time registered owner or tenant who has not benefited personally from a subsidy. Furthermore, beneficiaries must occupy the property and cannot transfer it for a period of five years. The report proposed that funds for categories one and two come from the public sector, while the private sector should play a direct role in the provision of finance and site development in categories three and four. The De Loor report also suggests the establishment of a housing finance corporation which would facilitate the provision, *inter alia*, of commercially based loans for the purchase of land, loans for bulk infrastructure and rental housing, short-term development finance and funds for upgrading and urban renewal.

Finally, proposals are made for a new institutional framework, which would include a single non-racial ministry of housing with jurisdiction over all of South Africa (including the ten 'homelands'). It is also recommended that the housing functions of other central government institutions be rationalized and located in the proposed housing finance corporation (South Africa, 1992).

The UF has criticized the De Loor proposals for marginalizing the poor, being 'antagonistic' to the market approach and recommending an 'inappropriate' role for the government. The UF argued that approximately 60 per cent of unaccommodated households would fall into category one of the proposed subsidy policy and would enter a dependency relationship with the state in that such households are not granted title and are tenants of the state. The UF emphasized that while site ownership was not essential, the dependency cycle should not be sustained. The UF points out that the report fails to give adequate attention to the consolidation process and to mechanisms which will protect the interests of poor communities regarding the identification and acquisition of well-located land for settlement (Urban Foundation, 1992b).

The UF further maintains that despite the De Loor proposals' adherence to market principles, the proposals result in 75 per cent of unaccommodated households (categories one and two of the subsidy proposals) having to depend on the government for housing provision. Consequently, it is argued by the UF that the public sector would become the main player, rather than a regulator, in housing provision, implying that local and regional government would need to increase in size with increased delivery capacity. Furthermore, the proposed subsidy approach is criticized as too complex, thus necessitating additional administration (Urban Foundation, 1992b).

PLANACT has welcomed aspects of the De Loor proposals but criticized others. It welcomed the suggestion that there be a synergistic relationship between the government and the housing market with the government's role being to steer the market toward socially desirable goals. It regards the proposal as a shift from official housing policy in the 1980s, where the government divested itself of responsibility for providing formal shelter for the poor. For PLANACT, another positive aspect is the suggestion to establish a negotiation mechanism for identifying land and establishing townships, as land-use planning and township establishment were until recently unilaterally undertaken by the central government (PLANACT Housing Policy Team, 1992).

However, PLANACT contends that the assumption that there is a 'capital shortage' in South Africa is false. Capital is deemed to exist in the financial markets, insurance companies and pension funds, but has largely not been directed to financing housing development projects (Bond, 1992, 1994). PLANACT further argues that the report's macrofinancing proposals are weak as they do not consider the potential dangers of deregulated housing finance and dependency on foreign capital. The assumption that people below the breadline can afford deposits, service housing loans and service charges is also questionable. The report's subsidy proposals also fail to target consolidating informal housing or to consider differential land costs. Moreover, PLANACT points out that there is no provision for consolidating the serviced site into a serviced formal house (PLANACT Housing Policy Team, 1992).

Common to mainstream agencies such as the Urban Foundation, the Independent Development Trust, the South African Housing Trust and the De Loor proposals of the South African government is their support for a broadly defined market-oriented approach to the resolution of South Africa's crisis of shelter provision. This is not to suggest that all such organizations' policies and strategies are similar in all respects: certain agencies place more emphasis on delivering housing (product-focused), while others have been attempting to rectify the imbalance by directing resources to programmes which build the capacities and skills of those meant to benefit from housing schemes. Contrary to the intention of most mainstream housing policy formulators, alternative approaches aim to build capacities of communities to enable agents in shelter delivery and consolidation, and to gain a measure of control over development processes affecting their lives. Attention now turns to these alternative, process-focused, approaches.

ALTERNATIVE HOUSING POLICIES

Although South Africa, unlike many Third World nations, has a well-developed financial infrastructure and private sector, a relatively sophisticated system of housing finance and delivery is no guarantee that the accommodation needs of all households will be met. It is, therefore, of import to confront a critical issue: as the international experience suggests (cf. Forrest, 1987; Pugh, 1989; Robben and van Stuijvenberg, 1986) private ownership and control of land and property, as implied in all mainstream solutions, cannot assure housing for all. Should South African housing policy formulators and planners, therefore, not look beyond the private market, and explore housing production, management and delivery systems which do not operate according to the logic of profit? Two broad stances to non-commodified housing production and management systems can be discerned: an anti-statist, neo-populist approach, of which housing credit co-operatives are the best example, and a social-democratic, neo-corporatist approach where the focus is on state-directed public-housing programmes.

Credit co-operatives, retail lending institutions and housing provision

Housing co-operatives have been advanced as an alternative to commodified housing production and management systems (Chouinard, 1989). It is also estimated that the annual income of savings clubs and burial societies in South Africa amounts to over R600 million (*Business Day*, 27 September 1990), suggesting a potential capital pool into which co-operative housing schemes can tap. Savings clubs are non-profit co-operatives which not only save and lend money, but are also subject to democratic control by members. In some instances, members contribute an equal amount of cash per month, and at a predetermined date money and accumulated interest are shared equally. Another example would be cash contributions by members on a monthly basis, but with one specific member entitled to the entire money pool for that month. The procedure is repeated until all members have had an opportunity to draw from the pool, after which the cycle is repeated.

An example is the Cape Credit Union League. New members pay a joining fee of R100 and start with savings by purchasing at least two shares worth R1 each. Cash is deposited in a bank, members obtain 1 per cent interest per month. The problem is simply that the amounts saved by individuals in such schemes are too small to make a significant contribution to housing. A more successful housing credit co-operative is the Group Credit Company operated by the UF. In 1989 the UF set up the Group Credit Company as a loan facility for the informal housing sub-market, which formal financial institutions are not currently serving. Regarding actual housing delivery, toward the close of the 1970s the UF established five housing utility companies (non-profit organizations) to provide housing to low-income people. Over 26 000 serviced sites and 25 300 formal low-cost houses have been delivered over the past decade.

Another innovation was the establishment, together with the ULA Merchant Bank and the Urban Foundation, of a funding scheme: the collateralized housing investment paper (Chips) aimed at raising R500 million for financing low-cost housing through group loans. The scheme was established to attract investment from private financial institutions who would normally not invest in low-cost housing. Investors were to be provided with a Chip, backed by a portfolio of fixed-interest rate securities and other external securities which would be managed by the ULA Merchant Bank. Risks were to be minimized by insuring a fixed-interest rate of 16.5 per cent on loans, by spreading funds across a wide base of small loans to different borrowers (loans do not exceed R12 500), and by providing group loans (Pretorius, 1990). The IDT Finance Corporation has also supplied loan finance to some 12 retail lending institutions (RLI) located countrywide which aim to provide loan funding to the poor to acquire shelter and establish business ventures. Most RLIs provide loans of between R500 and R10 000 to people earning around R1500 a month. RLI's grant loans to individuals and groups.

With a *stokvel* (group credit society), funds are lent to a group of people who provide the deposit for the purchase of a house and take responsibility for the repayment of the loan, each person drawing the required amount from the loan. If a person were to default on repayment of his or her share of the loan, the entire group would lose its deposit. Group pressure ensures that loans are honoured. Examples of RLIs include the Amatola Finance Corporation, Bochabela Finance Corporation and the Eastern Transvaal Group Credit Company. The success or otherwise of RLIs is still uncertain given the newness of these initiatives; at the time of writing many institutions had yet to begin operating (cf. *Financial Mail*, 21 May 1993; *Housing in Southern Africa*, April 1993). Given the uncertainty surrounding neo-populist solutions, local policy formulators may benefit by examining successful state-directed public-housing programmes.

Public housing revisited

Advocates of a South African public-housing programme confront a persistent, widespread and irresistible ideological prejudice against any serious housing programme directed by the state. Moreover, this ideological bias against public housing is shared across the spectrum: ranging from corporate think-tanks like the UF, to non-governmental development agencies such as PLANACT. In general, there appear three main, and rather spurious, assertions made against a state-directed public-housing programme.

First as Oelofse, senior manager of the Urban Foundation's Housing Policy Unit claimed, those Third World states, such as Hong Kong and Singapore, which have provided most of the citizenry with public housing do not present viable models transferable to the local situation (*Business Day*, 13 January 1994). It is asserted that the Singapore public-housing experience for example, is too idiosyncratic, and therefore irrelevant to South Africa. Nevertheless, in a critique of Oelofse's argument Lupton points out that although these two East Asian city-states are in many ways unique, what is at issue are those cross-cultural policy instruments and planning mechanisms which are transferable: an efficient technocracy; stable fiscal policies; long-term planning (*Business Day*, 24 January 1994). Above all, what is required is the requisite political will to embark on a sustained public-housing programme (*Business Day*, 24 January 1994). Indeed, the utility of the Hong Kong and Singapore experiences is not the specificity of these two societies but those aspects of their public-housing programmes which are transferable (Castells *et al.*, 1991). Housing planners and policy formulators could therefore do themselves a service by carefully examining the Hong Kong and Singapore models of state-directed housing provision.

A second, and more pernicious dismissal pertains to the assumption that a country requires economic growth before it can afford a sustained public-housing policy. Once more, the evidence from the Far East contradicts the claims of local policy-makers (cf. Castells *et al.*, 1991). In short, state-directed public-housing programmes actually served as the backbone of some of the most spectacular paths of sustained economic growth in recent years. Indeed, in Singapore and Hong Kong, two of the Orient's economic miracles, public-housing provision actually *preceded* economic take-off and was *functional* to the remarkable growth in GDP achieved by both countries (Castells *et al.*, 1991).

A final set of criticisms relates to the social costs incurred in the East Asian experiences. Undoubtedly, Singapore and Hong Kong are not paragons of Western-style liberal democracy. Singapore has an authoritarian, neo-racist regime while colonial rule is still the order of the day in Hong Kong. The point is, however, that there is no causal relationship between politically authoritarian or repressive regimes on the one hand and economic growth and public housing on the other (Castells *et al.*, 1991). On the contrary, as Castells *et al.* (1991) point out, many Third World authoritarian states have failed dismally in the realms of economic development and public housing. Despite obvious examples where public housing policies have been successful both in meeting the accommodation needs of most citizens, and contributing to sustained economic growth, South African housing policy appears to be heading in another direction: proliferating squatter settlements within a crisis-ridden economy.

CONCLUSION: TOWARDS A POST-APARTHEID NATIONAL HOUSING POLICY

To conclude provisionally, the discussion now turns to exploring the contours of a future national housing policy. In particular, the National Housing Forum, a body representing the most coherent attempt to formulate a viable housing policy and strategy will be examined. Thereafter, some general aspects which may comprise an actual future national housing policy will be suggested.

By the early 1990s it was widely recognized that South Africa had no coherent and legitimate national housing policy. Various arms of the state responded in an *ad hoc* and fragmentary manner to burgeoning squatter settlements in urban centres, sometimes turning a blind eye, and at other times relocating residents of spontaneously established informal settlements to provincial administration-subsidized site-and-service schemes situated on the peripheries of urban centres, thereby perpetuating the inequitable and inefficient structure of the spatially fractured apartheid city. Also, the provision of low-cost housing in well-located residential areas had been reduced to a trickle. Besides, by mid-1990 migrant worker hostels on the central Witwatersrand had become key elements in the upsurge of violence essentially between township residents and hostel residents, coinciding with the launching of Inkatha Freedom Party branches in the Transvaal.

In response to requests from the Kagiso Trust and the National Interim Civic Committee which later became the South African National Civic Organisation (SANCO), the IDT and the Development Bank of Southern Africa (DBSA) convened a meeting to discuss the crisis around migrant worker hostels. Various stakeholders in housing were present at the meeting where it was agreed that the hostels question had to be conceptualized as part of a broader housing problem which could be resolved within an all-encompassing development framework. The National Housing Forum was formally launched in August 1992, a year after the first meeting to examine the hostels question. The long-term aim of the forum is to negotiate for a workable and non-racial future direction for the housing sector in South Africa with a particular emphasis on the provision of housing to members of disadvantaged communities. While the Forum will not actually provide housing itself, it will provide the forum where broadly supported practical plans and policies are formulated to address the country's housing crisis.

Given the demands for Africanization, the Forum is regarded as the most widely 'representative' negotiation structure yet established to examine housing in the country. Forum membership includes certain political parties (African National Congress, the Azanian People's Organization, the Inkatha Freedom Party and the Pan Africanist Congress), black civic organizations (SANCO), building material supply and construction companies, financial institutions, employer organizations, trade unions and development organizations (DBSA, IDT, Kagiso Trust, SAHT, and the UF) (*National Housing Forum News*, Number 1, Spring 1992). While the government was initially a participant in the Forum, it withdrew before it was formally launched arguing that it considered the Forum to be an interim government by stealth in the housing arena. The government nevertheless maintained contact with the Forum and submitted a copy of the De Loor report to it for discussion (*Finance Week*, 7 February 1992), and during 1993 the government and the Forum conducted negotiations on a range of substantive issues. The strength of the Forum is that for the first time non-governmental organizations with a stake in housing are directly participating in housing policy formulation. In particular, black nationalist organizations such as the African National Congress and certain civic organizations are now active Forum participants. There is also an understanding that acceptable housing solutions must encompass both technical and political factors. Previous official housing task teams sought technocratic solutions to simplifying the delivery of houses to passive consumers of apartheid policies. At the Forum, however, the emphasis is on participation and negotiation of all interest groups with a stake in housing in the formulation of an acceptable housing policy.

Furthermore, in recognizing that certain interest groups have been disadvantaged by apartheid, the Forum is conscious of the need to level the playing fields. To this end programmes have been implemented to build the capacity of disadvantaged groups effectively to identify the needs and priorities of their constituencies and actively to contribute to Forum proceedings (*National Housing Forum News*, Number 2, Summer 1992). The Forum has divided its work into three phases. The first phase involved the compilation and review of existing data on the relevant issues, split among six working groups examining the following: land and services; end-user finance and subsidies; housing types and delivery systems; institutional structures, roles and fund mobilization; integrating the cities; and hostels. By the second half of 1993, the Forum was in its second phase of work, which involved exploring long-term policy options aimed at rationalizing and restructuring the housing sector and negotiations with the government around a joint housing initiative between the Forum and the government.

The joint initiative involves the spending of some R500 million set aside from the government's 1993–94 budget for low-cost housing to be directed through a non-racial structure controlled jointly by the Forum and the government. The Forum and the government had also reached agreement on the need to exercise joint control over the government-funded hostels upgrading programme. Regarding the latter initiative, it is anticipated that joint control will ensure that government funds are used efficiently and that all interest groups affected by hostel upgrading participate fully and effectively in decisions concerning the future of hostels. It is anticipated that phase three will involve the finalization of detailed, long-term policy options, which were formulated during the second phase, to have an immediate effect on the 1994–95 national budget, and to create an environment in which the government of national unity (transitional government) can operate effectively with a broadly supported housing policy and strategy alternative to the current apartheid housing and urban planning policies.

During negotiations the National Housing Forum and the government jointly concluded that an overall agreement on housing, including some key economic and political issues that would affect housing, was necessary. The agreement would ensure that housing be dealt with in the broader, national framework and that policies and strategies emanating from such a framework are likely to be sustainable (National Housing Forum, 1993). Considering the National Housing Forum's leverage, in that it represents all the main interests with a stake in housing in South Africa, it is likely that the government of national unity will take its recommendations seriously. In fact, it has been suggested that the Forum, together with the other forums dealing with other socio-economic issues, become permanent features of South Africa after the 1994 elections (Pillay and Richer, 1993), creating the space for a range of organizations of civil society to involve themselves in the reconstruction process and ensuring that policy reflects a range of interests.

Participants in the National Housing Forum are particularly conscious of the need to provide an environment in which both communities and individuals are able to develop their capacities to participate and to ensure that housing projects are sustainable. Linked to an emphasis on 'process' and community empowerment in development is the consolidation process, which may receive more attention in the future. Site-and-service schemes and *in situ* upgrading of informal settlements are likely to form a substantial portion of a future housing stock. However, various mechanisms are likely to be introduced to enable the general socio-economic upliftment of such environments on an incremental basis, rather than the state granting a once-off subsidy to poor and marginal urban residents and paying lip service to consolidation. A proportion of the budget may be set aside for consolidation purposes to ensure that people are not condemned to shacks without a reliable source of income in a poorly serviced environment.

A future housing policy is also likely to give greater attention to ensuring the integration of South Africa's spatially fragmented cities, which are both extremely inefficient and inequitable, particularly for the poor located on the urban peripheries. Mechanisms may be introduced to ensure that vacant and under-utilized urban land close to city centres is made available for housing, particularly low-cost housing (cf. Tomlinson, 1990). This will enhance the poor's access to employment opportunities and urban amenities and reduce infrastructural costs. Capital subsidies for the poor wishing to reside in inner-city areas may also be introduced. Such subsidies would provide the poor with more options regarding residential location than are now available.

Alongside making available for residential development underutilized and vacant urban land, measures may be introduced to increase the supply of urban land for the poor. Measures could include the removal of zoning constraints and building codes or the zoning of land for low-cost housing purposes which would increase land supply and reduce the per unit price (Smit, 1992). Market-related financial packages, which make small loans available to the poor without collateral such as the Group Credit Company, the Home Loan Guarantee Fund and the IDT-funded retail-lending institutions, are likely to continue to be supported in the future.

A proportion of funds set aside for housing from central government may also be used as collateral to entice financial institutions to lend further down-market. Also, prototypes not yet experienced in South Africa such as community-based banking and credit schemes are likely to be nurtured, together with housing co-operatives, section 21 companies as housing vehicles and community-controlled land trusts.

Finally, a future national housing policy is most likely to take shape around a broadly defined market-oriented framework, such as the De Loor proposals, albeit with certain important modifications. Policy planners are likely to be particularly aware of the need to ensure a sustainable housing strategy that does not jeopardize macroeconomic stability. Moreover, an essential element of future housing policy will be a focus on 'process' as opposed to product delivery. The challenge is to seek a healthy balance between these two equally important elements of housing provision. At the end of the day, the dominant South African housing policies will be market-oriented, as alternative approaches such as public-housing programmes and decommodified types of shelter provision have been effectively sidelined in urban policy debates.

REFERENCES

Bond, P. (1990) 'Township housing and South Africa's "financial explosion": the theory and practice of finance capital in Alexandra', *Urban Forum*, 1 (2), 39–67.
Bond, P. (1992) 'De Loor Report is off the mark', *Work in Progress*, 84, September, 2–4.
Bond, P. (1994) 'Aspects of wholesale and retail housing finance for a democratic South Africa', unpublished paper presented to the AIC/Euromoney Conference on Financing Low Cost Housing: Setting Guidelines, Policies and Initiatives for the Future, Johannesburg, 16 January.
Castells, M., Goh, L. and Kwok, Y-W. (1991) *The Shek Kip Mei Syndrome: Economic Development and Public Housing in Hong Kong and Singapore*, Pion, London.
Chouinard, V. (1989) 'Social reproduction and housing alternatives: co-op housing in post-

war Canada', in M. Dear (ed.), *The Power of Geography*, Unwin Hyman, Boston, 222–37.

Cullinan, K. (1993a) 'Face to face with the Independent Development Trust', *Work in Progress: Reconstruct Supplement*, March, 2–3.

Cullinan, K. (1993b) 'Transforming the Independent Development Trust', *Work in Progress: Reconstruct Supplement*, January/February, 1–2.

Denoon, D. (1973) *A Grand Illusion: the Failure of Imperial Policy in the Transvaal Colony During the Period of Reconstruction*, Oxford University Press, London.

Department of Local Government and National Housing (1992) *Annual report*, Department of Local Government and National Housing, Government Printer, Pretoria.

Forrest, R. (1987) 'Spatial mobility, tenure mobility, and emerging social divisions in the UK housing market, *Environment and Planning A*, 19, 1611–30.

Hindson, D. (1987) *Pass Controls and the Urban African Proletariat*, Ravan, Johannesburg.

Horrell, M. (1978) *Laws Affecting Race Relations in South Africa, 1948–1976*, South African Institute of Race Relations, Johannesburg.

Lupton, M. (1993) 'Collective consumption and urban segregation in South Africa', *Antipode*, 25 (1), 32–50.

Mabin, A. and Parnell, S. (1983) 'Recommodification and working-class home ownership: new directions for South African cities?' *South African Geographical Journal*, 65 (2), 148–66.

National Housing Forum (1993) Unpublished paper prepared for the PWV Regional workshop, August 1993, Johannesburg.

Oelofse, M. and van Gass, C. (1992) 'End-user finance and Subsidies', unpublished paper presented to the National Housing Forum, Working Group 2, December.

Pillay, D. and Richer, P. (1993) 'Coordination of forums', *Work in Progress*, July/August, 2–4.

PLANACT Housing Policy Team (1992) 'De Loor from a developmental perspective', *Work in Progress*, October 1992, 6–7.

Posel, D. (1992) *The Making of Apartheid*, Clarendon Press, Oxford.

Pretorius, F. (1990) 'Explanatory memorandum on the Urban Foundation and Mortgage Lenders Association's Loan Guarantee Initiative', unpublished paper, Independent Development Trust, Johannesburg.

Pretorius, F. (1990) 'Chips funding for low-cost housing', unpublished paper, Independent Development Trust, Johannesburg.

Pugh, C. (1989) 'Housing policy reform in Madras and the World Bank', *Third World Planning Review*, 11 (3), 249–73.

Robben, P. and van Stuijvenberg, P. (1986) 'India's urban housing crisis: why the World Bank's sites and service schemes are not reaching the poor in Madras, *Third World Planning Review*, 8 (4), 335–45.

Smit, D. (1992) 'Housing policy options: the De Loor proposals and an alternative', unpublished paper presented to the Economic Trends Workshop, Cape Town, 5–7 June.

South Africa (1992) *Housing in South Africa: Proposals on a Policy and Strategy*, report prepared by the Task Group on National Housing Policy and Strategy, South African Housing Advisory Council, R.P. 79–1992 (Chairman: J. De Loor), Government Printer, Pretoria.

South African Housing Trust (SAHT) (1992) *Annual Report*, SAHT, Johannesburg.

SAIRR, 1980, *Survey of Race Relations 1979*, South African Institute of Race Relations, Johannesburg.

SAIRR, 1987, *Race Relations Survey 1986*, South African Institute of Race Relations, Johannesburg.

SAIRR, 1990, *Race Relations Survey 1989/90*, South African Institute of Race Relations, Johannesburg.
SAIRR, 1992, *Race Relations Survey 1991/92*, South African Institute of Race Relations, Johannesburg.
SAIRR, 1994, *Race Relations Survey 1993/94*, South African Institute of Race Relations, Johannesburg.
Swilling, M. (1990) 'Deracialised urbanisation', *Urban Forum*, 1 (2), 1–38.
Tomlinson, R. (1990) *Urbanization in Post-apartheid South Africa*, Unwin Hyman, London.
Tomlinson, R. (1992) 'Competing urban agendas in South Africa', *Urban Forum*, 3, 92–105.
Urban Foundation (1990) *Housing for All: Proposals for a National Urban Housing Policy, Urban Debate 2010: 9*, Urban Foundation, Johannesburg.
Urban Foundation (1992a) *Annual Review*, Urban Foundation, Johannesburg.
Urban Foundation (1992b) 'Notes on the De Loor proposals', unpublished report, July.
Walker, N. and Merrifield, A. (1993), 'Overview of housing delivery systems in South Africa', summary status report presented to the National Housing Forum, working group 3, February.
World Bank (1992) *Aide Memoire: South Africa Urban Reconnaissance Mission*, World Bank, Washington DC.

9 The Employment Challenge in a Democratic South Africa

CHRISTIAN M. ROGERSON

INTRODUCTION

Questions surrounding job creation, unemployment and poverty alleviation must be high on the policy agenda of decision-makers in a democratic South Africa. Debates concerning unemployment—its size, nature and causes—garnered increased attention especially from the 1970s as reflected in a growing body of research, academic writings and government reports (Urban Foundation, 1987; Thomas, 1990). In particular, a rising temperature of controversy and concern related to employment matters has been a noticeable feature of the 1980s and early 1990s. Recent survey findings taken across all segments of South African society identify unemployment or the need for employment opportunities as the foremost challenge facing policy-makers (Ligthelm and Kritzinger-Van Niekerk, 1990, p. 629). Accordingly, it is not surprising that employment questions occupy a major part of the work undertaken by new institutional structures of decision-making such as the National Economic Forum (Patel, 1993) or by the rash of regional development forums which have emerged across South Africa since 1992 (Bekker and Humphries, 1993; Rogerson, 1994a).

Job creation was identified as a policy matter demanding 'immediate attention' by the National Economic Forum, a consensus-geared co-operative body which was launched in October 1992 to bring together representatives from organized labour, government and big business (Patel, 1993). At the sub-national and local levels, new organizations are surfacing across South Africa seeking to grapple with unemployment issues; for example, the Natal–KwaZulu Job Creation and Enterprise Development Initiative was formed in February 1993 with the explicit purpose of designing and implementing a regional job creation strategy to enhance local employment options (Natal–KwaZulu Job Creation and Enterprise Development Initiative, 1993, p. 2). Likewise, in the deliberations and research of other regional development forums, such as those for the Border–Kei region, the PWV or the Western Cape, issues relating to employment needs are highly prominent (Border–Kei Development Forum, 1992; Hirsch, 1992a; Mabin and Hunter, 1993). Finally, the need to address the imperatives of employment creation has been central to several proposals and programmes for rethinking directions of national economic development planning in a new post-apartheid South Africa.

The Geography of Change in South Africa, edited by Anthony Lemon.
© 1995 by the Editor and Contributors. Published in 1995 by John Wiley & Sons Ltd.

The objective in this chapter is to weave together some of the most important research and key areas of contemporary debate concerning the employment challenge that confronts South African decision-makers in the 1990s and beyond. The chapter unfolds in three sections of discussion. First, the magnitude of South Africa's employment problems is examined in terms of documenting the deteriorating job situation in the formal sector of the economy. Issues to be highlighted include the stagnation or shrinkage in job creation in key employment sectors such as agriculture, manufacturing and mining, and the factors underpinning the diminished absorptive capacity of the formal sector. Against this backdrop, in the second section an examination is pursued of several policy proposals advanced to ameliorate the mounting crisis of unemployment. The focus is upon the creation of possible income opportunities across the formal as well as the informal sectors of the South African economy. It is argued that a distinction needs to be made between immediate needs and emergency programmes for work creation and longer-term objectives of restructuring the development path of a democratic South Africa. In the concluding section, a number of comments are offered on the regional dimension of South Africa's employment problems. In particular, the discussion spotlights key implications of the withering-away of the top-down central state-led regional development programme and of the accompanying rise of a host of local level initiatives for economic development. The overall thrust and direction followed in the chapter is therefore to shift from initially examining structural considerations of unemployment and policies for job creation to interpreting aspects of the broad geography of employment change.

SOUTH AFRICA'S WORSENING UNEMPLOYMENT CRISIS: DIMENSIONS AND CAUSES

The extent and deterioration of South Africa's employment dilemma is disclosed through a series of studies conducted by the Development Bank of Southern Africa (DBSA) and its associates (Ligthelm and Kritzinger-van Niekerk, 1990; Coetzee et al., 1991; DBSA, 1991a, 1991b, 1993; Coetzee, 1992). This research provides a base for interpreting the shrinkage of the country's formal economy and the accompanying rise of unemployment.

THE REDUCED ABSORPTIVE CAPACITY OF THE FORMAL ECONOMY

The current overall picture is that of a potential South African workforce, estimated at 16.8 million in 1991, of which almost nine million or 53.5 per cent were without formal employment opportunities. Those populations outside formal wage labour either were unemployed or working in the country's informal or subsistence economies. Although the statistical information concerning under-employment in the informal economy or 'subsistence' activities (primarily rural agriculture) is contentious, it is generally accepted that these two spheres of activity 'cannot provide a socially acceptable minimum level of living for a large proportion of those who are currently engaged in them' (DBSA, 1993, p. 2). Accordingly, with a deepening of the labour market crisis and a national workforce expanding at a rate of 2.7 per cent annually, the impact on human welfare has been disastrous as mirrored in escalating levels of poverty, violence and socio-political instability.

From Table 9.1 it is evident that since the mid-1960s South Africa's unem-ployment problem has worsened to an alarming extent as a consequence of the declining absorption capacity of the country's formal economic sector. Whereas during the 1960s the share of the labour force without formal employment fluctuated around 30–3 per cent, growing to 35 per cent by 1976, since 1980 the number of people without formal wage work has escalated considerably, reaching over 50 per cent of the national workforce by 1990. Moreover, on the basis of projections made by DBSA (1993), less than 43 per cent of South Africa's potential labour force will be in formal wage employment by 1995.

Table 9.2 provides further evidence of the decreased capacity of South Africa's formal economy to absorb new workseekers. The diminished labour absorption capacity of the economy is measured by relating the average annual increase in the workforce to average annual increase in formal employment (Coetzee et al., 1991, p. 181; DBSA, 1991a, 1991b). The DBSA research shows over a 30-year period a dramatic decline in the absorptive capacity of South Africa's economy (see Table 9.2). Specifically, this research reveals that the potential labour force (or total labour supply) increased between 1960 and 1990 at an average annual rate of 2.9 per cent from 6.9 million (1960) to 16.3 million (1990). Consequently, the number of new entrants to the labour market per annum between 1985 and 1990 (392 600) was more than double the number of new entrants annually between 1960 and 1965 (194 800). By contrast, the formal economy was able only to generate new formal employment opportunities at an average annual growth rate of 1.8 per cent from 1960 to 1990 (representing an increase from 4.6 million opportunities in 1960 to 7.9 million by 1990). Moreover, the average annual increase (incremental) in formal employment opportunities between 1985 and 1990 (33 000) was less than one-quarter of the average annual increase in the period 1960–65 (157 600) (DBSA, 1991a). Put another way, annual net additions to the formal sector decreased from 81 per cent in the early 1960s to only 8.4 per cent by the late 1980s; this implies that of more than 1000 new entrants per day to the labour market, only 84 could be absorbed into formal sector work (DBSA, 1993, p. 2). The serious economic recession that afflicted South Africa during the period 1989 to end-1991 erased all employment gains made in the formal economy in the years 1986 to 1989. Accordingly, in the five-year period 1986–91 when the labour force increased by a factor of 2.4 million people, the net increase

Table 9.1. South Africa: Potential Labour Force, Formal Employment and Outside Formal Employment, 1960–91

Year	Potential Labour Force[1] ('000)	Formal Employment Opportunities[2] ('000)	(%)	Without Formal Employment Opportunities[3] ('000)	(%)
1960	6901	4652	67.4	2249	32.6
1961	7065	4852	68.5	2233	31.5
1962	7275	4961	68.2	2314	31.8
1963	7470	5012	67.1	2458	32.9
1964	7669	5190	67.7	2479	32.3
1965	7875	5440	69.1	2435	30.9
1966	8085	5608	69.4	2477	30.6
1967	8302	5724	68.9	2578	31.1
1968	6524	5845	68.6	2679	31.4
1969	8752	6023	68.8	2729	31.2
1970	8985	6164	68.6	2821	31.4
1971	9283	6269	67.5	3014	32.5
1972	9590	6326	66.0	3264	34.0
1973	9908	6597	66.6	3311	33.4
1974	10 236	6509	66.5	3427	33.5
1975	10 576	6942	65.6	3634	34.4
1976	10 926	7078	64.8	3845	35.2
1977	11 288	7145	63.3	4143	36.7
1978	11 662	7176	61.5	4486	38.5
1979	12 049	7298	60.6	4751	39.4
1980	12 453	7450	59.8	5003	40.2
1981	12 830	7649	59.6	5181	40.4
1982	13 200	7803	59.1	5397	40.9
1983	13 582	7757	57.1	5825	42.9
1984	13 975	7832	56.0	6143	44.0
1985	14 377	7788	54.2	6589	45.8
1986	14 760	7798	52.8	6962	47.2
1987	15 140	7858	51.9	7282	48.1
1988	15 530	7958	51.2	7572	48.8
1989	15 930	7987	50.1	7943	49.9
1990	16 340	7953	48.7	8387	51.3
1991	16 790	7799	46.5	8991	53.5
1995[4]	18 571	7962	42.9	10 509	57.1

1. Consists of 95% of the males and 55% of the females in the 15–64 age group.
2. Total number of formal employment opportunities available.
3. Potential labour force minus formal employment opportunities.
4. The estimated average annual growth of 0.51% in formal employment assumes a growth of 2% and 0.74% per annum respectively in GDP and fixed capital stock.
Source: DBSA, 1993.

Table 9.2. South Africa: Absorption Capacity of the Formal Economy, 1960–91

Period	Average Annual[1]	Annual Incremental[2]
1960–65	68.1	81.0
1965–70	68.9	65.3
1970–75	66.4	49.1
1975–80	62.0	27.2
1980–85	57.2	17.7
1985–90	51.0	8.4
1985–91[3]	50.2	0.5
1990–95	44.5	0.4

1. The average annual absorption capacity for the period.
2. The ability of the formal economy to absorb the net number of new entrants, annually, into the labour market.
3. Resulting from a decline of 154 000 employment opportunities from 1990 to 1991.
Source: DBSA, 1993.

in formal employment opportunities was a dismal 12 000 jobs (DBSA, 1993, p. 2).

The inability of the formal economy to absorb new workseekers varies with different degrees of severity across the different regions of South Africa (DBSA, 1991b). Using the DBSA's definition of development regions, the absorptive capacity of the country's formal economy on a regional basis varied from 22.4 per cent to 58.6 per cent in 1990. Those development regions which recorded the lowest absorption capacities in 1990 were Region E (Natal–KwaZulu and Northern Transkei) at 35.2 per cent, Region D (Eastern Cape, Ciskei and Southern Transkei) at 33.7 per cent, and Region G (Northern Transvaal, Lebowa, Gazankulu and Venda) at 22.4 per cent (DBSA, 1991a, 1991b). These three regions exhibit poor absorptive capacities despite high rates of temporary out-migration of a large segment of their potential labour to other regions in an effort to secure formal employment. None the less, it must be noted that while the major regions of attraction for migrant workers did record better labour absorption capacities—namely, Region A (Western Cape) at 57.4 per cent and Region H (PWV region) at 58.6 per cent—these regions were also unable to supply adequate formal employment opportunities to their own de facto labour forces as well as the migrant workers from other development regions (DBSA, 1991a).

At the root of the declining absorptive capacity of the South African economy, particularly during the 1980s, is the disappointing employment performance recorded by the sectors of agriculture, mining and manufacturing. In absolute terms, this decade witnessed, inter alia, a stagnation in employment levels in the manufacturing sector with 1990 job levels (approx 1.5 million jobs) no higher than the 1982 peak (Levy, 1992; Black, 1993); a net downturn in employment in the mining sector with the shedding of some 50 000 jobs between 1980 and 1990 (from 766 600 jobs in 1980 to 713 800 by 1990) largely as a result of the low gold price since 1986; and a precipitate decline in agricultural employment from a total workforce of 915 700 in 1980 to an estimated 761 200 by 1990 (DBSA, 1991a, p. 21). Indeed, for the period 1980–90, it is apparent that the relative importance of the agricultural, mining and manufacturing sectors as employment generators has decreased in significance; correspondingly, sectors such as finance and business services and construction have strengthened in importance. DBSA (1991a, p. 20) data reveal that in 1980 agriculture contributed more than 12 per cent to total employment with the contributions of the mining and manufacturing sectors at 10.3 per cent and 19.6 per cent respectively as compared with 9.6 per cent, 9.0 per cent and 19.1 per cent for the three sectors in 1990 respectively. By contrast, the contribution of finance and business services strengthened from 3.9 per cent in 1980 to 5.6 per cent in 1990 and that of the construction sector from 5.4 per cent in 1980 to 5.9 per cent in 1990.

THE SCOPE AND CAUSES OF UNEMPLOYMENT

The weakened capacity of the formal sector of the South African economy to match the high growth rate of the potential labour force inevitably results in a growth of what the DBSA (1991a, 1991b) calls the 'peripheral sector' (a term which covers involvement in both the informal economy and the subsistence agricultural sector) and, more importantly, of open unemployment. Admittedly, data concerning levels of unemployment in South Africa must be treated with some caution due to problems with both official definitions and means of measuring unemployment (Thomas, 1990); estimates of unemployment differ widely and no single acceptable figure really exists (Coetzee, 1992). Nevertheless, irrespective of statistical source, a persistent upward trajectory in rates of national unemployment is recorded since the mid-1960s (Ligthelm and Kritzinger-van Niekerk, 1990, p. 630). The best assessment on the recent condition of unemployment is presented by DBSA data which show that the rate of unemployment increased markedly from 18.4 per cent or 1.7 million people in 1980 to reach 36.9 per cent or 4.7 million people by 1991 (Table 9.3). An even gloomier picture, however, is painted by data released in 1993 by the South African Reserve Bank which suggest that rates of national unemployment are perhaps as high as 46 per cent (*Sunday Times* (Johannesburg), 29 August 1993).

In terms of the geographical distribution of unemployment, significant variations exist across the country. DBSA research showed that in 1990 the unemployment rate in South Africa's major metropolitan areas was approximately 14 per cent. Between the metropolitan centres the lowest levels of 10.8 per cent unemployed were recorded for the Cape Peninsula; the worst problems of unemployment were experienced in the depressed Port Elizabeth region where

Table 9.3. Unemployment including Estimates for Hidden Unemployment,
1980, 1985 and 1991[1]

Year	Number of People Unemployed ('000 000)	Unemployed as Percentage of Economically Actice Population[2] (%)
1980	1.7	18.4
1985	3.2	29.2
1991[3]	4.7	36.9

1. Hidden unemployment represents those persons who are active in the peripheral sectors of the economy due to the fact that they are unemployed. They are still actively seeking formal employment.
2. The economically active population consists of all workers whether employer, employee, self-employed or unemployed.
3. Unadjusted population census figures.
Source: DBSA, 1993.

there was a rate of 25 per cent unemployment. That the metropolitan centres shared an average level of unemployment considerably above that of 9.5 per cent indicated for South Africa's small towns and white rural areas (that is non-homeland regions) is indicative of the impact of migration on these regions' unemployment problems. One study suggested that 'at least three workers arrive in urban areas for every one job being created' (Ligthelm and Kritzinger-Van Niekerk, 1990, p. 632). Nevertheless, the most severe problems of unemployment remain concentrated geographically in the bantustan regions where unemployment rates were conservatively recorded as 35 per cent (DBSA, 1991a).

In racial terms, the burden of unemployment is especially acute among South Africa's black population, who form almost 85 per cent of the total unemployed; none the less, a much publicized trend in the early 1990s is for the increasing spread of unemployment into the ranks of the white community. A further finding is that unemployment rates for females and young people are almost twice as high as rates for males and older workers (DBSA, 1993, p. 3). Overall, it is clear that the levels and volume of unemployment have consistently expanded over the past two decades, even during upswings in the business cycle, a finding which suggests that unemployment in South Africa assumes structural features 'implying it is related not only to the level but also to the pattern of economic activity in the country' (Ligthelm and Kritzinger-van Niekerk, 1990, p. 630).

Analyses of the causes of South Africa's weak employment performance point to a complex of underlying factors. These range from the negative impacts of macroeconomic strategies on employment creation to political complications, labour market inefficiencies, the actions of trade unions, low productivity of capital and labour, high population growth rates, lack of skills among a large segment of the population, sanctions and capital flight and institutional factors such as the overregulation of the economy (after Ligthelm and Kritzinger-van Niekerk, 1990; Thomas, 1990; DBSA, 1993). At the heart of the problem is, however, the falling rate of national economic growth since the mid-1960s which was inseparable from the broader crisis of the international economy in general and of semi-peripheral countries in particular (Vieira et al., 1992). Between 1960 and 1974 the national economy expanded at an average of 5.5 per cent annually but declined to a meagre 1.8 per cent for the period 1975–88 (Coetzee, 1992). During this period of emasculated growth rates, the rate of expansion of the national population consistently exceeded economic growth resulting in an erosion of per capita incomes of 1 per cent annually (Coetzee, 1992). Further aggravating the unemployment situation has been the nature of the growth process which has been capital-intensive rather than employment-intensive (DBSA, 1993, p. 3).

Research interpreting the disappointing employment performance in the sectors of agriculture, mining and manufacturing points to the common theme of the South African economy experiencing a capital-intensive rather than a labour-intensive growth trajectory. James (1992, p. 22) notes a shift to more labour-efficient and productivity-centred mining operations and to a marked trend in gold mining, especially after the growing strengthening of trade union organization, for labour substitution through technological innovation. The march of capital-intensive production in the South African countryside is a theme advanced in Lipton's (1993) analysis of the agricultural sector. This analysis is confirmed by a recent World Bank (1993a, p. 16) investigation which points progressively to a capital-intensive pattern of agricultural development with mechanization, including heavy investments in tractors and combine harvesters, contributing to a major reduction in farm employment opportunities. Several studies show that, in large measure, the poor employment creation record of South Africa's manufacturing sector is attributable to high and rising levels of manufacturing capital intensity (Kaplinsky, 1992a, 1992b; Levy, 1992). In particular, this research points to the massive capital investments during the 1980s poured by the state into nurturing the build-up of certain strategic sectors, most notably of armaments manufactures and of the SASOL synfuels and MOSSGAS chemical ventures. Thus, it is ironical that one of the outcomes of the international sanctions campaign against South Africa was to push the economy on to a more capital-intensive growth path (Lipton and Simkins, 1993, p. 12).

NEW EMPLOYMENT INITIATIVES AND ECONOMIC RESTRUCTURING

In the light of the foregoing analysis, it is evident that the employment challenge for policy-makers in a democratic South Africa centres attention on a number of key issues. First, there is the short-term question of urgently providing some form of immediate relief in order to address the massive problems of the unemployed. A second and longer-term requirement is, however, to restructure the patterns of growth in order to enhance the employment-creating capacity of the economy. In particular, as argued by Lipton and Simkins (1993, p. 23), South African development strategy in the 1990s must aim to secure the twin goals of attaining a growth rate significantly in excess of the population growth rate of 2.3 per cent per annum and a 'reduction of poverty and inequality by means of fuller and more equitable participation in the economic and social spheres by historically disadvantaged groups', especially the black population. Finally, a set of issues which straddle the nature of development processes both of a short- and long-term nature concern the future role of (and policy formulation towards) the informal economy. Overall, as the DBSA (1993, p. 4) asserts, in South Africa 'a comprehensive and integrated set of policy measures is needed to significantly impact upon the unemployment problem' in an immediate as well as in a long-term manner.

SHORT-TERM JOB CREATION PROGRAMMES

Several observers and organizations have pressed (or are currently pressing) the case for the introduction of well-conceived and adequately scaled programmes for public works (PWPs) or other special employment programmes (SEPs) in South Africa (Abedian and Standish, 1986, 1989). The National Economic Forum is, for example, discussing schemes for a National Development Corps to furnish job opportunities for unemployed young people (Patel, 1993, p. 9). The issue of retrenchments early attracted the attention of COSATU to possibilities for emergency work creation programmes (Keet, 1991).

Indeed, as part of its proposed Programme for Reconstruction and Development, COSATU supports a programme for job creation through large-scale public works in order to kickstart the economy and extend social services to previously disadvantaged communities (Naidoo, 1993). Sketchy available details on this Roosevelt New Deal type of programme suggest that it might involve as many as 300 000 people with participants wearing uniforms and submitting themselves 'to moderate military-type discipline in exchange for three meals a day, subsistence wages and accommodation in tent towns or barracks' (Star (Johannesburg), 19 June 1993). Another SEP proposal emanating from the National Youth Development Forum is for a national youth service programme, which aims to target basic education, training and a small stipend on segments of South Africa's 'lost generation' of youths (Weekly Mail and Guardian (Johannesburg), 1–7 October 1993).

The unifying goal of all these initiatives would be to exert an immediate impact on South Africa's worsening problems of poverty and unemployment in a similar manner to that effected in the 1930s by the massive and successful public works programme launched against white poverty (Abedian and Standish, 1986, 1989). An essential prerequisite for enacting any such form of programme is a restructuring at the macroeconomic level of public-sector expenditure to favour the unemployed and poor (Ligthelm and Kritzinger-van Niekerk, 1990; DBSA, 1993). Support for this form of macroeconomic restructuring and of employment-creation through public works programmes has been signalled by the African National Congress (1993, p. 14).

One kind of SEP initiative which already has been implemented in South Africa relates to the experience of emergency drought relief measures in rural areas. Although the guidelines for drought relief projects provided a base for 'participatory development' to be led by the needs of communities, evaluations of these programmes have pointed to important weaknesses with regard to their poor productivity, lack of continuity and often weak management (Donaldson, 1993). The most important lessons to be gleaned from their experience relate, however, to problems associated with 'community' participation, decision-making and control (see Nedcor and Old Mutual Professional Economic Panel, 1993). In particular, it is essential not to romanticize notions of 'community' in South African development planning for there are many poor black communities which are often so deeply divided that 'community control' of development resources becomes a source of local conflict (p. 79). Accordingly, what is suggested is that where there exists a real danger of fragmented communities being thrown into added conflict situations by the introduction of SEPs or PWPs, then government should assume a greater measure of control.

Currently, the most popular proposals for initiating new public works initiatives relate to the construction sector and the provision of urgently needed social and productive infrastructure. The building of schools, clinics and the extension of electricity supplies are some suggested areas for public-works job-creation efforts (DBSA, 1993). In addition, many observers maintain that the huge backlogs in both urban and rural water supply and sanitation systems across South Africa might be addressed through a government-led public works scheme. Projections are that an annual investment programme of roughly R500 million could catalyse over 30 000 jobs while meeting the basic needs of at least 500 000 people annually by providing new infrastructure as well as upgrading other public services (Standing Committee on Water Supply and Sanitation, 1993, p. 7). Another set of parallel proposals concerns the building of bulk infrastructure, irrigation works, road construction and maintenance work (DBSA, 1993).

Throughout the discussions on public-works schemes, stress is laid upon the necessity for adopting labour-intensive construction projects (Phillips et al., 1992; Patel, 1993). The National Housing Forum, for example, takes the view that all development projects that fall within the scope of low-income shelter, including infrastructure provision as well as schools or clinics, should be 'executed so as to achieve the optimum substitution of labour for equipment' (National Housing Forum, 1993, p. 2). The advantages of using labour-based construction are to promote community participation, facilitate the local retention of capital expended on projects and to transfer skills to the local community (Phillips et al., 1992; National Housing Forum, 1993).

Two important issues that warrant further attention in a national programme for job creation relate to regional impacts and the precise institutional arrangements required for project implementation. One suggestion made by the DBSA (1993) is that national public works and employment creation interventions will need a dedicated institution to manage the process in the short to medium term. The Bank recommends the option of a central employment creation programme to promote but not directly to implement activities with project responsibility devolved to the 'local and regional level for implementation by public, private and non-government institutions within a multi-year programme' (DBSA, 1993, p. 25). In terms of the spatial allocation of short-term job creation projects, suggestions have been made by the National Economic Forum that they be targeted particularly 'to underdeveloped areas and areas of high unemployment and social tension' (Patel, 1993, p. 10).

UNEMPLOYMENT AND THE INFORMAL ECONOMY

It is clear that the informal economy must assume a significant role in addressing the employment dilemmas of South Africa in the next decade (World Bank, 1993b). A large body of recent research has sought to document the nature, growth and workings of South Africa's informal economy and to initiate debate on its potential for resolving problems of burgeoning unemployment and poverty.

Estimating the dimensions and growth of South Africa's informal economy is fraught with definitional problems. The best available counts suggest that at the close of the 1980s at least 30 per cent of the total South African labour force was engaged in some form of informal work (Coetzee et al., 1991; Kirsten, 1991). For the early 1990s, Swainson (1992, p. 3) argues the total numbers of people engaged in informal economic activities 'could amount to between three and five million' with blacks the overwhelming majority in racial composition. All recent surveys and research point to a phase of very rapid expansion currently taking place in the local informal economy, especially in peri-urban zones and the mushrooming zones of informal settlements that girdle South Africa's major metropolitan centres (Rogerson, 1992a; World Bank, 1993b).

None the less, the advance of the informal economy is widespread across all South Africa's cities, towns and rural areas. In urban areas the march of the informal economy is manifest in the appearance of a host of new street or pavement-centred activities (including flea markets, hawkers, taxi drivers, street barbers, car washers and prostitutes), the proliferation of home-based enterprises (child-minding, *spazas* or informal shops, shebeens, backyard or garage workshops, hairdressers) and a small number of increasingly formalized ventures located at fixed business premises (small-scale manufacturers, liquor taverns)

(Rogerson, 1992b). Most dramatically, perhaps, the rising importance of the informal economy is evidenced by the expansion of urban cultivation on peripheral vacant land in a manner similar to that observed in other African cities (May *et al.*, 1993; Rogerson, 1993a).

In rural South Africa the growth of the informal non-farm economy involves households participating in an extraordinarily diverse range of activities which span craft manufacture, brewing, brick-making, traditional healing, dagga cultivation, the collection of medicinal herbs, fuelwood collection, the production of building materials and involvement in palm wine tapping (Rogerson, 1986; McIntosh and Friedman, 1989; Cunningham, 1991a, 1991b; McIntosh, 1991a; Preston-Whyte and Nene, 1991). Throughout the South African informal economy women are the prime actors; one major study found that women operate almost two-thirds of all the country's informal enterprise (World Bank, 1993b, p. 16).

As is shown within a rich local literature, the South African informal economy is not a homogeneous entity; instead, it comprises a diverse mixture of activities, development opportunities and problems. One vector of complexity in the contemporary informal economy surrounds the simultaneous occurrence of processes of the formalization of informal enterprise and the informalization of formal enterprise (Rogerson, 1992a). The former process includes the outcomes of movement towards legislative deregulation and creation of a more favourable business environment for a range of informal activities. Examples would include street trading, the operation of mini-bus taxis, liquor selling (the shift from shebeen to tavern) and the transition of many backyard producers into formalized industrial premises. The process of the informalization of formal enterprise refers to situations in which larger business enterprises seek to by-pass regulations covering employment protection and security by establishing or linking production to small, informal ancillary enterprises on terms which make those who work within them particularly vulnerable to exploitation. The informalization of formal enterprise occurs particularly in the manufacturing sector and is a process which attracts a hostile reaction from the organized trade-union movement (Manning and Mashigo, 1993). Examples of this informalization process have been documented in a number of 'splinter' operations taking place at the network of deregulated industrial parks and industrial hives which the Small Business Development Corporation has established over the past decade (Rogerson, 1991a).

In terms of development possibilities, important distinctions must be drawn across the continuum of enterprises from on the one hand a group of marginally profitable survival-oriented households and on the other a group of dynamic or potentially dynamic firms (World Bank, 1993b). This divide is sometimes made between the informal economy of bare survival and an informal economy of growth (Rogerson, 1991b). In between these polar positions, however, are many situations in which informal enterprise may link (through for example subcontracting arrangements) in subordinate fashion to institutions in the formal economy, creating simultaneously both opportunities for, and constraints on, future growth (Rogerson, 1991a).

The mass of informal enterprises in South Africa's large cities, towns and rural areas fall into the category of survival enterprises. Typical examples are the activities of urban cultivators, of groups of child-minders, of rural collectors of traditional herbs and of the ubiquitous township retail operations of *spazas* and hawkers. These survival enterprises are operated mainly by women entrepreneurs 'whose main motivation is to generate income rather than to grow' or alternatively are businesses run by individuals forced into self-employment by the weak formal labour market (World Bank, 1993b, p. 1). While incomes obtained in these enterprises are low, they represent important contributions to household incomes, even though they may not be the principal source of income. Essentially, such survival enterprises function as a welfare safety net to those populations forced to endure a miserable existence in what Booth (1987, p. 15) calls the 'dungeons' of the South African informal economy. A last sub-group of survival informal enterprise would encompass the host of urban and rural co-operative ventures (in handicrafts, sewing, construction or production) which surfaced during the past decade largely as defensive responses to a deteriorating economic climate. Several studies confirm that the record of South African co-ops has not been promising in terms of generating anything more than bare survival livelihoods for members (Etkind, 1989; Rogerson, 1990; Collins, 1992; Jaffee, 1992).

Although the various kinds of survival businesses are found in all geographical zones of the country, they seem to constitute the majority of the informal economy in peripheral settlements (Frankel, 1988; May and Stavrou, 1990), small towns (Davidson and Stacey, 1988; Nduna, 1993), and rural areas (McIntosh, 1991b; Preston-Whyte and Nene, 1991; Swainson, 1992). Policies to assist the informal economy of survival would be measures of a welfare-assistance kind and would necessarily vary between, for example, the specific needs of garbage scavengers, spaza retailers, pavement vendors, urban subsistence cultivators or child-minders (see Rogerson, 1991b, 1991c; 1993a). None the less, the package of policy measures would be geared at the level of individuals or households with the prime aim of reinforcing the safety-net function of this part of the South African informal economy as a last buffer against destitution (Rogerson, 1991b, p. 219). Suggested types of assistance programmes might include improvements in health delivery, expansion of educational opportunities, the improved provision of safe drinking water and adequate sanitation, schemes for upgrading housing or furnishing temporary access to land for the poorest and most vulnerable communities (Rogerson, 1991b; May et al., 1993).

In the context of pressures for black economic empowerment, policy interventions concerned with growth and long-term employment creation must address the constraints affecting those groups of dynamic enterprises which potentially form an informal economy of growth. Currently, the most promising growth segments of South Africa's informal economy appear to lie in promoting branches of small-scale informal manufacturing, construction and transport enterprises (see Liedholm and McPherson, 1991; Standish and Krafchik, 1991; Rogerson, 1993b; World Bank, 1993b). For these types of enterprise it is clear that moves made during the last decade towards deregulation are an essential first step to permit new growth. None the less, there is a substantial body of evidence which shows that deregulation is no longer the most important issue as a number of other key internal and external constraints have been discerned (Liedholm and McPherson, 1991; May and Schacter, 1992; Rogerson, 1993b). Today, the principal internal constraints are a direct legacy of apartheid and include 'lack of adequate technical, administrative and managerial skills and lack of access to financing'; important external constraints are highly competitive markets, lack of market infrastructure, and an often violence-racked business environment (World Bank, 1993b, p. 53). Finally, acknowledgement must be made also of the constraints placed on informal enterprise development arising from South Africa's pattern of dispersed urban growth (Dewar and Watson, 1991; Behrens and Watson, 1992).

In terms of 'growing' the informal economy, much interest now centres on the prospects for catalysing the expansion of small-scale industry (Rogerson, 1991a, 1993c, 1993d; Manning and Mashigo, 1993) and construction enterprise (Krafchik, 1990; Krafchik and Leiman, 1991; Standish and Krafchik, 1991; World Bank, 1993b). In both sectors of production and construction, considerable attention centres on the prospects and initiatives for expanding subcontracting linkages between large- and small-scale or informal enterprise (Krafchik and Leiman, 1991; Rogerson, 1993d, 1993e). Small-scale construction enterprises are targeted as potential beneficiaries of a programme to kickstart the national economy through major programmes for infrastructure upgrading and mass shelter provision in urban townships and informal settlements (Krafchik, 1990; Standish and Krafchik, 1991). Manning (1993, p. 14) argues that there 'is a good chance that the informal manufacturing sector will grow, providing the poor with cheap clothing, shoes, furniture and so on' but cautions that future performance 'may be closely tied to the influx of cheap imports from China and other East Asian countries' which 'may flood the markets and make it very tough for informal and small-scale manufacturers to compete'. Overall, it is concluded that local 'informal manufacturers will have to identify market niches which are not being serviced by large manufacturers or imports' (p. 14). Although limited prospects exist for assisting the growth of some small-scale industrial and construction ventures in non-metropolitan places of South Africa, by far the most promising zones of opportunity in terms of markets and prospects for subcontracting to larger enterprise would seem to lie in and around the largest metropolitan centres of the PWV, Cape Town and Durban.

The World Bank's research on small micro-enterprise development in 'disadvantaged sections' of the South African community emphasizes a need for a major post-apartheid initiative to 'focus on building contract linkages between the formal private sector and qualified black businesses' (World Bank, 1993b, p. 51). The Bank advocates that a system of incentives be applied to encourage such linkages (most notably, subcontracting) and that larger enterprises should assume a 'mentoring' role with respect to black small-scale enterprise, furnishing both technical advice and training in order to achieve quality standards and procurement arrangements that would be voluntary and mutually beneficial (World Bank, 1993b, p. xvi). A further set of interventions are needed to strengthen linkages to the public sector, including possible changes in tendering systems to expand government procurement (especially at the local level) from black micro-enterprise. Another promising policy direction to stimulate an informal economy of growth in South Africa involves a break away from policies which are premised upon helping the individual small entrepreneur *per se*. Instead, the policy focus should be upon assisting the building and extension of *networks* of interacting and co-operative small enterprises. Policy initiatives and thinking on these lines draws inspiration from the international experience of successful networking among small enterprises in the local industrial districts of Italy, Denmark or southern Germany (Rogerson, 1993c, 1993d, 1993e).

LONG-TERM ECONOMIC RESTRUCTURING

In the first years of a new democratic government the immediate need will be to implement a measure of emergency short-term unemployment programmes and a sound package of policies to manage the diversity of South Africa's informal economy. The major task, however, will be to reconstruct the South African economy towards a long-term sustainable, more equitable and employment-creating growth path. This task of economic reconstruction will have to be confronted in an inauspicious global economic climate amid considerable competition from other semi-peripheral countries, especially those in East Asia (Vieira *et al.*, 1992, p. 195). Although precise details are unclear, the outlines of a path for a long-term restructuring of the South African economy are beginning to emerge. The broad directions for economic reconstruction are indicated in the works of influential institutions such as the National Economic Forum (Patel, 1993) or the DBSA (1993) and from policy documents issued by the ANC (1993), the major political actor in a new democratic dispensation.

The key principles for guiding a growth and development path for a post-apartheid South Africa have been enunciated by the ANC and its allies. The central goal of economic policy will be to mould a strong, dynamic and balanced economy that is directed towards, *inter alia*, eliminating the legacies of apartheid poverty and inequality, democratizing the economy, creating productive work opportunities for all South Africans at a living wage, initiating growth to improve the quality of life especially for the poor and fostering a 'balanced' regional economy (ANC, 1993, p. 12). To achieve these objectives the ANC proposes a two-pronged national economic strategy. First, is a set of redistribution pro-grammes to meet the basic needs of the poor. Second, are programmes for long-term restructuring 'on the basis of new, comprehensive and sustainable growth and development strategies in all sectors of the economy' (ANC, 1993, p. 12). The specific nature and changing course of South African development strategy, as guided by the key principle of 'democratizing economic growth' (Gelb, 1990), is examined now for two key sectors, namely agriculture and industry.

Agriculture

In redressing the gross maldistribution of land and other resources in favour of white farming, the ANC (1993, p. 17) supports a programme of land reform and agricultural restructuring designed to 'facilitate a move away from exclusive reliance on large-scale single crop agriculture, to a more diversified combination of agricultural production systems, including family farms, small scale farms and cooperative farming systems'. It is widely accepted that rural development, especially agricultural development, will have 'substantial implications for growth, job creation, and poverty reduction' (Lipton and Simkins, 1993, p. 23). The potential and options for land reform in a post-apartheid South Africa have been widely debated (World Bank, 1993a; Lipton, 1993). In terms of employment, it is essential 'that agriculture and the rural non-farming sector not only maintain their present labour force, but significantly increase the number of employment opportunities they provide' (Lipton, 1993, p. 379). Programmes for land reform, COSATU argues, should be viewed not only as a necessary element of redis-tribution 'but also as a growth-generating activity' (Naidoo, 1993). Policy changes are required 'to counter past policies that discouraged labour intensifi-cation' (DBSA, 1993, p. 15). Moreover, added scope for employment opportu-nities arises from potentially strengthening linkages both within the agricultural sector and to other economic sectors, including a push to expand opportunities for agri-business by adding value to primary agricultural products (DBSA, 1993, p. 15).

At the heart of employment debates are prospects for more labour-intensive farming both by independent smallholders and employers on large commercial farms. Lipton (1993, p. 378) maintains that 'there appears to be significant potential for labour-intensive smallholders in some sectors of South African agriculture'. None the less, it must be understood that there are major regional variations in potential for a viable smallholder sector; geographically, promising areas have been identified in parts of Natal, the eastern Transvaal lowveld, in the Western Cape and even in pockets of land within and around the overcrowded but sometimes underfarmed bantustan regions (Bromberger and Antonie, 1993; Lipton, 1993). The DBSA (1993, p. 16) is enthusiastic about the employment-creation prospects in bantustan regions from implementing small-farmer support programmes (FSPs) which would allow farmers to extend and intensify agricultural operations; projections are that in these blighted regions FSPs could furnish subsistence employment to an additional 100 000 households. Another possibility for fostering smallholder expansion is through developing market gardening in the periurban zones of South Africa's major metropolitan areas (May et al., 1993). In the case of small towns the key issue is to reconstruct small-scale agriculture in order to meet the needs of local urban markets. Dewar (1993) urges that the reconstruction of South Africa's network of small towns is important for restructuring the countryside through integration and the forging of powerful symbiotic relationships.

As several observers stress, however, there are significant sets of constraints which may impact upon future prospects for new job creation in South African agriculture and in rural development generally. Beyond the obvious (but none the less crucial) limitation on access to available land (DBSA, 1993), Lipton (1993, p. 382) directs attention to the question of the low wages and low incomes earned in farming; it is suggested that relatively high and rising formal sector wages in South Africa 'will make it difficult to stimulate labour-intensive agriculture'. As earnings of farmworkers are among the lowest in South Africa, a major challenge will be 'to combine the encouraging employment trends and possibilities with improvements in the earnings, conditions and status of agricultural workers' (Lipton, 1993, p. 380). Equally thorny is the issue of 'the willingness of South Africa's unemployed to work in agriculture' for the phenomenon has been observed that, despite escalating levels of unemployment, workers sometimes turn down agricultural employment opportunities due to low wages and often unpleasant, harsh working conditions. Although economic logic points to a need to accord priority to job creation and correspondingly to hold back increases in unskilled wages (especially when these result in labour-displacing capital investment), Lipton (1993) argues that it is likely that such a strategy would be opposed by the trade union movement.

Industry

Most policy observers see the reinvigoration of South Africa's stagnant manufacturing sector as the key sectoral issue in long-term economic development planning. Accordingly, the restructuring of South African industry and of associated policies (such as for foreign investment and trade) have been the subject of considerable research attention and debate (see Bell, 1990; Black, 1992a, 1992b, 1993; Kaplinsky, 1992a, 1992b; Levy, 1992; Joffe et al., 1993). Especially influential in informing directions for a new industrial strategy is likely to be the Industrial Strategy Project, initiated by COSATU and the Economic Trends Research Group, which is 'investigating the competitive status of 13 of South Africa's industrial sectors, assessing not only their abilities to withstand global competition, but also their potential for meeting basic needs' (Joffe et al., 1993, p. 2). A vital issue will be to restructure industry to a path of greater labour absorption, a challenge that might require a structural change from the presently highly concentrated pattern of control to a more diffuse structure of control of industry as well as growing labour-intensity particularly in those manufactures geared to supply basic human needs such as food, clothing or infrastructure (Levy, 1992, p. 50).

Although no consensus exists at present, certain key themes are emerging on the contested terrain of South African industrial policy. Black (1992b) points out that the debates about trade orientation—whether it should be inward-looking or export-oriented—overlook the fact that South African manufacturing cannot avoid the need for competition on world markets. Therefore, the choice 'facing future South African economic policy-makers is not *whether* the country will have to compete on world markets, but the nature of this competition and the types of products' (Black, 1992b, p. 140). A wide body of research suggests that South Africa would not be competing on international markets with newly industrializing countries on the basis of low wages. Rather, it is argued that the potential exists for the state to intervene to create new areas of comparative advantage by selective intervention and support (Black, 1992b; Joffe et al., 1993). Instead of cheap-labour manufactures much attention centres on the potential of South Africa's cheap energy and mineral wealth to furnish the basis of a large metals-based manufacturing sector (Joffe et al., 1993, p. 4). Great prospects appear to exist for South Africa to pursue a road of resource-based industrialization through expanding projects and developments linked to mineral beneficiation (Jourdan, 1992).

The experience of state intervention in East Asia in selective targeting of industrial sectors for state support is seen as providing a model for future South African policy-making (Black, 1992b). But instead of the state support formerly accorded to such strategic industries as armaments and chemicals during the apartheid years, Black (1993) argues for a coherent set of industrial policies that might furnish support to selected industries linked to the upgrading of productivity and the achievement of international competitiveness. A key institution in future restructuring is identified as the Industrial Development Corporation (IDC), which is seen to possess many of the features of a state-owned industrial bank. Suitably re-oriented, Black (1992a, p. 17) avers, the IDC 'could have an important future role in supporting and guiding more appropriate forms of industrial development'. Changing patterns of state sectoral intervention undoubtedly will have major implications. Importantly, the progressive shift of state financing away from the military machine and the armaments industry poses considerable problems as regards defence conversion in the 1990s due to the closure or rundown of many South African defence-production facilities (see Rogerson, 1993f).

The role of multinational investors in a future South African industrial economy is another matter of considerable controversy. ANC policy documents proclaim that the democratic state 'will welcome foreign investment, in accordance with our objectives for growth and development, and will adopt an open approach to the entry of foreign investment' (ANC, 1993, p. 15). This policy direction was confirmed in recommendations made by the National Economic Forum which stated that South Africa should specially seek to attract investors who would 'strengthen the country's growth potential and competitiveness in the long term' (National Economic Forum, 1993, p. 84). Among others, Hirsch (1992b) cautions, however, that foreign investment will not pour in automatically to a post-apartheid, democratic South Africa, especially given the competing attractions of East Asia, Eastern Europe or the former Soviet Union. Significantly, the removal of international sanctions has not been so far accompanied by a rush of new or returning foreign investment to South Africa. None the less, a slow trickle of new foreign investment into industry was recorded throughout 1993. Three notable examples include the return of US multinational pharmaceuticals company Proctor and Gamble, which exercised an option to buy back the controlling stake in a company which it sold to local investors after disinvesting in 1986 (*Star*, 5 November 1993); the arrival of Daiwoo, the Korean electronics conglomerate, which is joining up with Anglo-American Industrial Corporation to enter the South African television sector (*Engineering News*, 1 October 1993); and the entry of Fiat's commercial vehicle concern, Iveco, into South African production through a joint venture arrangement with a local concern (*Engineering News*, 29 October 1993). The politics of inward foreign investment in a post-apartheid South Africa will be complex in the light of prospects of foreign firms taking up business opportunities at precisely the moment that the black business establishment hoped to secure a real stake in the South African economy. Hirsch (1992b, p. 40) suggests that one way of depoliticizing the issue of foreign investment and of strengthening black business might be for foreign investment policy to encompass strategies designed to boost black business enterprises variously by joint ventures, partnerships, licensing or franchise agreements.

Finally, during the 1990s South African manufacturing will have to confront the opportunities offered by the global shift to flexible production and the importance of competing on grounds of quality as well as price. Questions relating to the prospects for flexible specialization are firmly on the industrial policy agenda (Rogerson, 1994b). In particular, concern for the development possibilities associated with flexible production (including the introduction of new technologies, quality circles or just-in-time production systems) is widespread in the group of economists and advisers linked to the ANC and its allies (Kaplan, 1991; Kaplinsky, 1991a, 1991b; Joffe *et al.*, 1993). Indeed, some researchers argue that a transition to flexible production systems is 'vital in reconstructing the South African economy' (Maree, 1991, p. 83). Great potential is attached to resuscitating the country's flagging manufacturing capacity through a move to flexibility, niche marketing, niche targeting and the adoption of new managerial philosophies which acknowledge labour 'as a resource rather than as a cost' (Kaplinsky, 1991a, p. 197). Indeed, the extent to which industry in South Africa moves towards adopting flexible work practices will be crucial to the future growth or stagnation of subcontracting and correspondingly to the development of segments of black manufacturing enterprise (Rogerson, 1991a, 1993b). Considerable interest centres on the potential of flexible manufacturing systems in South Africa which might incorporate expanded subcontracting linkages between large and (especially black-owned) small firms (Rogerson, 1993f). None the less, while some local manufacturers are experimenting with new flexible technologies and work practices, the volume of flexible production occurring is still at a relatively low level (Rogerson, 1994b). The importance of innovating flexible production is in opening up new possibilities for encouraging an alternative kind of industrial development model based on flexible networks of small-firm cooperation (Joffe et al., 1993; Rogerson, 1993c, 1994b). This would marry South African industrial strategy to policies for the micro-enterprise or informal economy.

CONCLUSION—THE GEOGRAPHY OF EMPLOYMENT CHANGE

The future geographies of employment change in a democratic South Africa will be the product of the effects of powerful spatial biases introduced by new sectoral policies oriented to support a new economic and political order. Undoubtedly, much new state finance is likely to be channelled to improving basic needs and redressing poverty. The geographies shaped by these implicit spatial policies will be tempered by the effects of explicit state interventions in the space economy, most notably by the effects of any regional policy programmes.

A major task for policy-makers will be to engineer a comprehensive national urban policy, incorporating proposals for all levels in the South African urban hierarchy and linked to a sound rural development policy, in order to ensure that excessive conflict is avoided between the effects of implicit and explicit interventions. Nevertheless, since the introduction in 1991 of a new industrial development programme, regional policy in South Africa has weakened in terms of affecting the overall geography of employment. The state has retreated from the former apartheid decentralization programme which sought to channel industry away from the metropolitan regions into the country's peripheral regions, especially the poverty-stricken bantustans. This phasing out of decentralization incentives has precipitated a movement towards the de-industrialization of certain bantustans, notably Transkei and Ciskei, where substantial falls in manufacturing employment have been recorded (Border–Kei Development Forum, 1992; Nel and Temple, 1992; Rogerson, 1994a).

The demise of regional policy correspondingly has raised the prospects for a greater focus of new industrial employment growth in and around South Africa's leading metropolitan regions (Rogerson, 1991d). Moreover, these regions also appear set to receive the major stimuli of sectoral programmes to kickstart the economy through urban infrastructural improvement. Further reinforcing this metropolitan focus of new employment growth has been the emergence of active urban initiatives to restructure and boost the economies of South Africa's leading cities, even including attempts to capture the Olympic Games in 2004 (Rogerson, 1994b). For the non-metropolitan places of South Africa the prospects for new employment growth are less promising and seemingly contingent upon pro-active local or community development initiatives (see Dewar, 1993; Nel, 1993, 1994).

ACKNOWLEDGEMENTS

Thanks are due to the Centre for Science Development, Pretoria for financial assistance in this research, to Roddy Payne for resource inputs and to Jayne Garside for critical comments.

REFERENCES

Abedian, I. and Standish, B. (1986) 'Public works programme in South Africa: coming to terms with reality', *Development Southern Africa*, 3 (2), 180–98.
Abedian, I. and Standish, B. (1989) *Job Creation and Economic Development in South Africa*, Report No. 10, HSRC Investigation into Manpower Issues, Pretoria.
ANC (African National Congress) (1993) *Ready to Govern: ANC Policy Guidelines for a Democratic South Africa Adopted at the National Conference 28–31 May 1992*, ANC, Johannesburg.
Behrens, R. and Watson, V. (1992) 'Inward development: solutions to urban sprawl', *Indicator South Africa*, 9(3), 51–4.
Bekker, S. and Humphries, R. (1993) 'Regional forums: new frontiers, new creatures', *Indicator South Africa*, 10(4), 19–22.
Bell, R.T. (1990) *The Prospects for Industrialisation in the New South Africa*, Rhodes University, Grahamstown.
Black, A. (1992a) *Current trends in South African Industrial Policy: Selective Intervention, Trade Orientation and Concessionary Industrial Finance*, Working Paper No. 9, Economic Trends Research Group, Development Policy Research Unit, University of Cape Town, Cape Town.
Black, A. (1992b) 'Industrial strategy: lessons from the newly industrialized countries' in I. Abedian and B. Standish (eds), *Economic Growth in South Africa: Selected Policy Issues*, Oxford University Press, Cape Town, 128–46.
Black, A. (1993) 'The role of the state in promoting industrialisation: selective intervention, trade orientation and concessionary industrial finance' in M. Lipton and C. Simkins (eds), *State and Market in Post Apartheid South Africa*, Witwatersrand University Press, Johannesburg, 203–34.
Booth, D. (1987) Measuring the 'success' of employment creation strategies in the apartheid state, unpublished paper, University of Natal, Durban.
Border–Kei Development Forum (1992) 'A new deal for Border–Kei region', submission on job creation to the National Economic Forum's Short Term Working Group.

Bromberger, N. and Antonie, F. (1993) 'Black small farmers in the Homelands: economic prospects and policies' in M. Lipton and C. Simkins (eds), *State and Market in Post Apartheid South Africa*, Witwatersrand University Press, Johannesburg, 409–49.

Coetzee, S.F. (1992) 'Evaluating short- and long-term solutions to employment creation in South Africa', submission on job creation to the National Economic Forum's Short Term Working Group.

Coetzee, S.F., Davies, G.B., Olivier, J.J. and de Coning, C. (1991) Urbanisation, economic development and employment creation in the new South Africa in E.P. Beukes, W.J. Davies, R.J.W. van der Kooy and L.A. van Wyk (eds), *Development, Employment and the New South Africa: Proceedings of the Biennial Conference of Papers Presented at the Development Society of Southern Africa Conference*, Development Society of Southern Africa, Innesdale, 179–92.

Collins, A. (1992) *Producer Co-operatives in the Northern Region of South Africa, 1978– 87*, Working Paper No. 77, University of Cape Town Southern African Labour and Development Research Unit, Cape Town.

Cunningham, A. (1991a) 'The herbal medicine trade: resource depletion and environmental management for a "hidden economy"' in E. Preston-Whyte and C. Rogerson (eds), *South Africa's Informal Economy*, Oxford University Press, Cape Town, 196–206.

Cunningham, A. (1991b) 'Lean pickings: palm wine tapping as a rural informal-sector activity' in E. Preston-Whyte and C. Rogerson (eds), *South Africa's Informal Economy*, Oxford University Press, Cape Town, 254–61.

Davidson, J.H. and Stacey, G.D. (1988) 'A potato a day from the pensioner's pay?: hawking at pension payout points', *Development Southern Africa*, 5 (2), 245–50.

DBSA (Development Bank of Southern Africa) (1991a) *Labour and Employment in South Africa: a Regional Profile*, Centre for Information Analysis, Development Bank of Southern Africa, Halfway House.

DBSA (Development Bank of Southern Africa) (1991b) *South Africa: an Inter-regional Profile*, Centre for Information Analysis, Development Bank of Southern Africa, Halfway House.

DBSA (Development Bank of Southern Africa) (1993) 'Employment creation strategies for South Africa', submission on job creation to the National Economic Forum's Short Term Working Group, Halfway House.

Dewar, D. (1993) Reconstructing the South African countryside: the case of the small towns, unpublished paper, University of Cape Town.

Dewar, D. and Watson, V. (1991) 'Urban planning and the informal sector' in E. Preston-Whyte and C. Rogerson (eds), *South Africa's Informal Economy*, Oxford University Press, Cape Town, 181–95.

Donaldson, A.R. (1993) 'Basic needs and social policy: the role of the state in education, health and welfare' in M. Lipton and C. Simkins (eds), *State and Market in Post apartheid South Africa*, Witwatersrand University Press, Johannesburg, 271–320.

Etkind, R. (1989) 'Co-operatives and socialism: some perspectives on co-operative management', *Transformation*, 9, 51–65.

Frankel, P. (1988) Urbanisation and informal settlement in the PWV complex, unpublished draft report for the Centre for Policy Studies, University of the Witwatersrand, Johannesburg.

Gelb, S. (1990) 'Democratising economic growth: alternative growth models for the future', *Transformation*, 12, 25–41.

Hirsch, A. (1992a) '"Growing the Cape": "establishment" and "liberation" forces in regional co-operation', *South African Labour Bulletin*, 16(5), 55–8.

Hirsch, A. (1992b) 'Inward foreign investment in a post-apartheid SA—some policy considerations', *South Africa International*, 23(1), 39–44.

Jaffee, G. (1992) 'Co-operative development in South Africa', in G. Moss and I. Obery (eds), *South African Review 6: from 'Red Friday' to Codesa*, Ravan, Johannesburg, 364–77.

James, W.G. (1992) *Our Precious Metal: African Labour in South Africa's Gold Industry, 1970–1990*, David Philip, Cape Town.

Joffe, A., Kaplan, D., Kaplinsky, R. and Lewis, D. (1993) Meeting the global challenge: a framework for industrial revival in South Africa, unpublished paper prepared for the IDASA meeting, South Africa's International Economic Relations in the 1990s, 27–30 April.

Jourdan, P. (1992) *Mineral Beneficiation: some Reflections on the Potential for Resource-based Industrialisation in South Africa*, Working Paper No. 14, Economic Trends Research Group, Development Policy Research Unit, University of Cape Town.

Kaplan, D. (1991) 'New technology: new skills, new opportunities?', *South African Labour Bulletin*, 15(8), 51–4.

Kaplinsky, R. (1991a) 'Technology and reconstruction colloquium: concluding remarks' in *Proceedings, Technology and Reconstruction Colloquium*, University of Cape Town Department of Adult Education and Extra-Mural Studies and Development Policy Research Unit, Cape Town, 185–99.

Kaplinsky, R. (1991b) 'A growth path for a post-apartheid South Africa', *Transformation* 16, 49–55.

Kaplinsky, R. (1992a) South African industrial performance and structure in a comparative context, unpublished paper prepared for the meeting of the South African industrial strategy project, Johannesburg, 6–10 July.

Kaplinsky, R. (1992b) Inter-sectoral determinants of capital intensity in South African manufacturing: have capital-intensive sectors been over-expanded?, unpublished paper prepared for the COSATU/Economic Trends Project on an industrial strategy for a post-apartheid South Africa.

Keet, D. (1991) 'Employment creation programmes: COSATU outlines proposals', *South African Labour Bulletin*, 16 (2), 42–7.

Kirsten, M. (1991) 'A quantitative assessment of the informal sector in E. Preston-Whyte and C. Rogerson (eds), *South Africa's Informal Economy*, Oxford University Press, Cape Town, 148–60.

Krafchik, W.A. (1990) Small scale enterprises, inward industrialization and housing: a case study of subcontractors in the Cape Peninsula low-cost housing industry, unpublished MA dissertation, University of Cape Town.

Krafchik, W. and Leiman, A. (1991) 'Inward industrialization and petty entrepreneurship: recent experience in the construction industry' in E. Preston-Whyte and C. Rogerson (eds), *South Africa's Informal Economy*, Oxford University Press, Cape Town, 345–64.

Levy, B. (1992) *How can South African Manufacturing Efficiently Create Employment?: an Analysis of the Impact of Trade and Industrial Policy*, World Bank Southern Africa Department Informal Discussion Papers on Aspects of the Economy of South Africa, Paper No. 1, Washington D.C.

Liedholm, C. and McPherson, M.A. (1991) *Small Scale Enterprises in Mamelodi and Kwazakhele Townships, South Africa: Survey Findings*, GEMINI Technical Report no. 16, Bethesda, Maryland.

Ligthelm, A. and Kritzinger-van Niekerk, L. (1990) 'Unemployment: the role of the public sector in increasing labour absorption capacity of the South African economy', *Development Southern Africa*, 7 (4), 629–42.

Lipton, M. (1993) 'Restructuring South African agriculture' in M. Lipton and C. Simkins (eds), *State and Market in Post Apartheid South Africa*, Witwatersrand University Press, Johannesburg, 339–407.

Lipton, M. and Simkins, C. (1993) 'Introduction' in M. Lipton and C. Simkins (eds), *State and Market in Post Apartheid South Africa*, Witwatersrand University Press, Johannesburg, 1–34.

Mabin, A. and Hunter, R. (1993) Report of the review of conditions and trends affecting development in the PWV, unpublished report prepared for the PWV Forum.

Manning, C. (1993) 'Dynamo or safety net: can the informal sector save the day?', *Work in Progress*, 87, 12–14.

Manning, C. and Mashigo, P. (1993) Manufacturing in micro-enterprises in South Africa, unpublished report to the Industrial Strategy Project, University of Cape Town.

Maree, J. (1991) 'Trade Unions and the democratisation of technology', *South African Labour Bulletin*, 16(2), 76–83.

May, J.D. and Schacter, M. (1992), 'Minding your own business: deregulation in the informal sector', *Indicator South Africa*, 10(1), 53–8.

May, J. and Stavrou, S. (1990) 'Surviving in shantytown: Durban's hidden economy', *Indicator South Africa*, 7(2), 43–8.

May, J., Attwood, H., Dominik, T., Kaye, B., Newton, N., Rogerson, C. and Witt, H. (1993) Development options for peri-urban agriculture, unpublished report prepared for the World Bank, Data Research Africa, Durban.

McIntosh, A. (1991a) 'Making the informal sector pay: rural entrepreneurs in KwaZulu' in E. Preston-Whyte and C. Rogerson (eds), *South Africa's Informal Economy*, Oxford University Press, Cape Town, 243–53.

McIntosh, A. (1991b) 'Rural income-generating activities: collective responses', in E. Preston-Whyte and C. Rogerson (eds), *South Africa's Informal Economy*, Oxford University Press, Cape Town, 279–89.

McIntosh, A. and Friedman, M. (1989) 'Women's producer groups in rural KwaZulu: limits and possibilities', *Development Southern Africa*, 6 (4), 438–53.

Naidoo, J. (1993) 'Planning the reconstruction of South Africa', *The Star*(Johannesburg), 13 July.

Natal–KwaZulu Job Creation and Enterprise Development Initiative (1993) ' "Windows of opportunity" for job creation in Natal/KwaZulu: a discussion paper', submission on job creation to the National Economic Forum's Short Term Working Group.

National Economic Forum (1993) 'Long-term working group report' in E. Patel (ed.), *Engine of Development?: South Africa's National Economic Forum*, Juta, Cape Town, 81–95.

National Housing Forum (1993) Technical submission to the National Economic Forum from the National Housing Forum on 'Housing and Job Creation Through Labour-Based Construction', submission on job creation to the National Economic Forum's Short Term Working Group.

Nduna, J. (1993) Street trading in Transkei's major urban areas, Unpublished MA dissertation, Department of Geography, University of the Witwatersrand, Johannesburg.

Nedcor and Old Mutual Professional Economic Panel (1993) 'Instituting public works programmes', submission on job creation to the National Economic Forum's Short Term Working Group (based upon the work of Nicoli Nattrass).

Nel, E. (1993) Regional economic development—a community perspective, unpublished paper, Rhodes University, Grahamstown.

Nel, E. (1994) 'Local development initiatives, a new development paradigm for urban areas?: a general assessment of this approach with reference to the experience of Stutterheim, *Development Southern Africa*, 11 (2), forthcoming.

Nel, E. and Temple, J. (1992) 'Industrial development and decentralisation in Transkei and the Border region', *Journal of Contemporary African Studies*, 11, 154–77.

Patel, E. (1993) 'New institutions of decision-making: the case of the National Economic Forum' in E. Patel (ed.), *Engine of Development?: South Africa's National Economic Forum*, Juta, Cape Town, 1–16.

Phillips, S., Meyer, D. and McCutcheon, R. (1992) 'Employment creation, poverty alleviation, and the provision of infrastructure: lessons from the labour-based construction of municipal services in Ilinge', *Urban forum*, 3(2), 81–113.

Preston-Whyte, E. and Nene, S. (1991) 'Black women and the rural informal sector' in E. Preston-Whyte and C. Rogerson (eds), *South Africa's Informal Economy*, Oxford University Press, Cape Town, 229–42.

Rogerson, C.M. (1986) 'Reviving old technology?: rural handicraft production in southern Africa', *Geoforum*, 17 (2), 173–85.

Rogerson, C.M. (1990) ' "Peoples factories": worker co-operatives in South Africa', *GeoJournal*, 22 (3), 285–92.

Rogerson, C.M. (1991a) 'Deregulation, subcontracting and the "(in)formalization" of small-scale manufacturing' in E. Preston-Whyte and C. Rogerson (eds), *South Africa's Informal Economy*, Oxford University Press, Cape Town, 365–85.

Rogerson, C.M. (1991b) 'Policies for South Africa's urban informal economy: lessons from the international experience' in E. Preston-Whyte and C. Rogerson (eds), *South Africa's Informal Economy*, Oxford University Press, Cape Town, 207–22.

Rogerson, C.M. (1991c) 'Home-based enterprises of the urban poor: the case of spazas' in E. Preston-Whyte and C. Rogerson (eds), *South Africa's Informal Economy*, Oxford University Press, Cape Town, 336–44.

Rogerson, C.M. (1991d) 'Beyond racial Fordism: restructuring industry in the "new" South Africa', *Tijdschrift voor Economische en Sociale Geografie*, 82 (5), 355–66.

Rogerson, C.M. (1992a) 'Tracking the urban informal economy', in G. Moss and I. Obery (eds), *South African Review 6: from 'Red Friday' to Codesa*, Ravan, Johannesburg, 378–87.

Rogerson, C.M. (1992b) 'The absorptive capacity of the informal sector in the South African city' in D.M. Smith (ed.), *The Apartheid City and Beyond: Urbanization and Social Change in South Africa*, Routledge, London, 161–71.

Rogerson, C.M. (1993a) 'Urban agriculture in South Africa: scope, issues and potential', *GeoJournal*, 30 (1), 21–8.

Rogerson, C.M. (1993b) 'Re-balancing racial economic power in South Africa: the development of black small-scale enterprise', *GeoJournal*, 30 (1), 63–72.

Rogerson, C.M. (1993c) 'Industrial districts: Italian experience, South African policy lessons', *Urban Forum* 4 (2), 37–53.

Rogerson, C.M. (1993d) 'Industrial subcontracting and home-work in South Africa: policy issues from the international experience', *Africa Insight*, 23 (1), 47–54.

Rogerson, C.M. (1993e) 'Looking to the Pacific Rim: production sub-contracting and small-scale industry in South Africa', paper presented to the International Geographical Union Conference on Industrial Space Meeting, Tokyo, August.

Rogerson, C.M. (1993f) Defence conversion: international experience, some South African policy issues, unpublished report prepared for the PWV Economic and Development Forum, Johannesburg.

Rogerson, C.M. (1994a) 'South Africa: from regional policy to local economic development initiatives', *Geography*, 79, in press.

Rogerson, C.M. (1994b) 'Flexible production in the developing world: the case of South Africa', *Geoforum*, 25, in press.

Standing Committee on Water Supply and Sanitation (1993) 'Job creation in the water supply and sanitation sector', submission on job creation to the National Economic Forum's Short Term Working Group.

Standish, B. and Krafchik, W. (1991) 'Inward industrialisation, subcontracting and the construction sector in the Western Cape', *Development Southern Africa*, 8, 203–14.

Swainson, N. (1992) Formalising the informal: training for the informal sector in South Africa, unpublished paper for the NEPI Conference 'Human resources policy for a new South Africa', Durban 7–10 May.

Thomas, W. (1990) 'Unemployment and the job creation challenge' in R.A. Schrire (ed), *Critical Choices for South Africa: an Agenda for the 1990s*, Oxford University Press, Cape Town, 250–73.

Urban Foundation (1987) Report on employment, unpublished report, Private Sector Council on Urbanisation – Working Group on Employment and Regional Development, Johannesburg.

Vieira, S., Martin, W.G. and Wallerstein, I. (eds) (1992) *How Fast the Wind?: Southern Africa, 1975–2000*, Africa World Press, Trenton, New Jersey.

World Bank (1993a) South African agriculture: structure, performance and implications for the future, unpublished report of the Agriculture and Environment Division, Southern Africa Department, World Bank, Washington DC.

World Bank (1993b) Characteristics of and constraints facing black businesses in South Africa: survey results, unpublished draft paper presented at the seminar on 'The Development of Small and Medium Business Enterprises in Economically Disadvantaged Sections of the South African Communities', Johannesburg 1–2 June.

Part III

SOUTH AFRICA IN
SOUTHERN AFRICA

10 South Africa and SADC(C): a Critical Evaluation of Future Development Scenarios

EDWARD RAMSAMY

INTRODUCTION

The rapidly changing international political economy has necessitated a redefinition of foreign policy objectives of nation-states as well as a re-evaluation of national development agendas. These changes pose new challenges and dilemmas for regional co-operation and inter-state relations. Tostensen (1993) points out that the southern African region has experienced major changes in the eighties and nineties. These include a greater emphasis on economic liberalization, democratization of political institutions, a reduction in international development assistance for the region and the prospect of a new order in South Africa. In response to some of these changes, the Southern African Development Co-ordination Conference (SADCC) reconstituted itself as the Southern African Development Community (SADC) in a summit meeting in August 1992. One of the major objectives of this change was to restructure the organization from a loose association of states to a more effective instrument for regional integration. The significance and regional implications of SADCC's restructuring are explored in greater detail in Chapter 11 of this volume.

The aim of this chapter is to examine economic and political relationships between South Africa and the member states of SADC in the context of current politico-economic restructuring. It will be argued that relations between South Africa and the other states in the southern African region will depend upon how domestic political actors in each state view regional co-operation and co-ordination. In this respect, the chapter challenges the basic premises of the dependency paradigm, which views the southern African regional political economy solely from the perspective of South African hegemony. The basic assumption of the dependency paradigm is that the internal political and economic crises experienced by SADC member states are attributable to the external, destabilizing effects of South Africa and geopolitically significant powers like the United States and the former Soviet Union. Rather, it is argued that intra-/inter-state problems in southern Africa resulted from a complex interplay of domestic and international factors. While a post-apartheid South Africa could potentially play an important role in southern Africa's future, the ability of South Africa to influence positively

The Geography of Change in South Africa, edited by Anthony Lemon.
© 1995 by the Editor and Contributors. Published in 1995 by John Wiley & Sons Ltd.

the development process in the region will depend upon South Africa's own political and economic fortunes in a post-apartheid era and the way in which domestic crises are resolved within other southern African states.

The arguments and observations in the chapter are divided into three parts. The first section proposes a theoretical framework for conceptualizing regional alliances and dependent/interdependent relations in southern Africa. Second, the chapter offers an overview of the structure and objectives of the SADCC and critically examines its successes and failures. Finally, the complexities associated with South Africa's future involvement in the SADC will be examined and the different possibilities for regional co-operation in a post-apartheid era will be explored.

NATION-STATES AND THE INTERNATIONAL SYSTEM: IN SEARCH OF A THEORETICAL FRAMEWORK

Over the past few decades, theories of international development in the Third World have been dominated by the modernization/diffusionist and the dependency paradigms. The conceptual flaws inherent in these theoretical constructs are aptly summarized by Lipietz (1987, p. 5).

> Despite the undeniable formal superiority of the imperialism-dependency approach, it seems that, like the rival liberal approach, [the stages of economic development] it had degenerated into an ahistoric dogmatism . . . It is as though the two theories were contemplating the development of history . . . if the South was stagnating, one theorist would tell you precisely what time it was; if new industrialization was taking place, another would say I told you so . . .

Both the dependency and modernization frameworks were based on a sense of finalism. Pre-specified outcomes were associated with the development process with little or no attempt to examine the spatial and temporal contingencies of development. Furthermore, Haggard (1990, p. 21) asserts that both these perspectives neglect politics in their analyses and tend to 'ignore how domestic political forces constrain economic policy and shape state responses to the external environment.' One way to approach this dilemma is to consider the advice given by Lipietz (1987, p. 3) that we 'obviously have to take into account the historical and national diversity of capital accumulation in each nation state under consideration'. Recently, several attempts have been made to correct these shortcomings in theoretical formulations of development, by 'bringing the state back in' as a key explanatory variable in social analysis (Evans et al., 1985; Jessop, 1990; Shafer, 1994). Instead of viewing the international system as the determinant of national development, the comparative political economy perspective (Evans and Stephens, 1988) takes a more recursive view by arguing that the global political economy simultaneously shapes and is shaped by the historical trajectories of development within individual nation-states. According to Haggard, this perspective aims to answer the following questions:

- Why are different development strategies chosen?
- Why do they persist?
- Why do they shift?

This view holds that external dependence and imperialism can be grasped only after the structure, function and policies of the state and domestic political economy are understood. While national development takes place in an international context, there is a variety of possible combinations of national policies that are available for local political actors to pursue. Thus, according to Gourevitch (1986, p. 64), one ought to take the following factors into consideration in attempting to ascertain specific directional shifts in state policies and alliances:

What is the position of the country with respect to the world economy?

Within the society, whom does the policy benefit? Who supports it? Who opposes it? Does the policy correspond to the wishes of a significant coalition of interests?

Who defines the policy alternatives, both in terms of debate and adoption? Is the policy formulated outside or within the state apparatus?

How is the policy legitimated and who constitutes the opposition to policy? How effective is the state in imposing a policy? What kinds of opposition are possible?

It is within the context of these issues that I conceptualize the turbulent political economy of southern African states. While advocates of the modernization approach tend to overlook the political and economic constraints faced by southern African states in formulating effective development strategies, dependency theorists, on the other hand, seem to 'liquidate the unique history and development of specific countries' (Milkman, 1979, p. 262). An understanding of SADC member states' relationships with South Africa, both in terms of past trends and future prospects, can be enhanced if the conceptual framework acknowledges how South Africa's dominance has impeded developmental efforts in southern Africa, while simultaneously grappling with the internal crises and contradictions in the national development agendas of the various states in southern Africa.

THE SOUTHERN AFRICAN DEVELOPMENT CO-ORDINATION CONFERENCE (SADCC)

AIMS AND OBJECTIVES OF SADCC

The Southern African Development Co-ordination Conference was established in 1980, consisting of nine southern African states: Angola, Botswana, Lesotho,

Malawi, Mozambique, Swaziland, Tanzania, Zambia and Zimbabwe (see Fig. 11.1). As a consequence of the demise of white minority rule in southern Africa, it was envisaged that the formation of a regional co-ordinating committee would offer the first 'real' opportunity for political co-operation and economic development in the region (Thompson, 1986; Meyns, 1984). The Lusaka Declaration of 1980 established four principal goals:

- (1) The reduction of economic dependence particularly, but not only, on the Republic of South Africa.
- (2) The forging of links to create genuine and equitable regional integration.
- (3) The mobilization of resources to promote the implementation of national, interstate and regional policies.
- (4) Concerted action to secure international co-operation within the strategy for economic liberation.

The basic aim in the formation of SADCC was the establishment of collective self-reliance. In essence, it was envisaged that SADCC member states would generate higher levels of trade among member countries. SADCC was also concerned with 'dependence-reduction' (Blumenfeld, 1991) but also attempted to avoid 'traditional' approaches to regional economic co-operation in light of the failures of such attempts elsewhere. No central authority was established to make central plans for the region; SADCC operated for three years without a central secretariat. The current SADC secretariat is located in Gaborone, Botswana; decision-making is integrated into the offices of the national state executives of the region. Thus, SADCC policy prescriptions were set by the cabinet ministers of the member states. Decision-making and policy formulation took place at an annual meeting of the nine heads of state during which each member state co-ordinated individual development sectors. Projects were neither carried out nor disbursed through SADCC. While the various sectoral commissions developed and presented to the donors a programme of project proposals, projects primarily remained the responsibility of the country in which the projects were located. The country in which the projects were to be implemented received funding from foreign donors; projects that were not country-specific were shared by the respective countries concerned (MacDonald, 1990).

The structure of SADCC represented a pragmatic, incremental approach to regional cooperation. Blumenfeld (1991, p. 138) appropriately concludes that the pragmatic philosophy adopted in the formation of SADCC was 'a matter of necessity' since its member countries were politically and economically diverse, with conflicting interests. Membership ranged from Botswana, whose political and economic system is based on free markets and liberal trade, to that of Angola and Mozambique, which initially espoused Marxist–Leninist ideologies of centrally planned economies based on Eastern European and Soviet models.

ACHIEVEMENTS OF SADCC

SADCC has been regarded as 'one of Africa's few successful attempts at regional cooperation' (Morna, 1990a, p. 49). Despite the broader structural limitations faced by SADCC countries, their endeavours were partly successful in attaining some of the goals established in the Lusaka Declaration. SADCC has concentrated on specific projects, thereby avoiding some of the problems encountered by other attempts at regional integration, especially the disproportionate burden placed on smaller and poorer states whose domestically produced goods are unable to compete with those of neighbouring developed states. Haarlov (1988, p. 176) appropriately concludes that:

> The specific approach of SADCC—distinct from traditional types of regional co-operation and integration—has been of paramount importance for the relative coherence, political backing and practical results until now.

SADCC's early concentration on the transportation sector has met with partial success. By early 1991, Mozambique's Beira corridor port was handling about two million tons of cargo annually. After current rehabilitation work is completed, the capacity of the port is expected to rise to five million tons annually. Sixty per cent of transit traffic from the six land-locked states was moving through SADCC ports in 1991, as compared with 20 per cent in 1980. SADC member states are now connected directly by satellite telecommunications, which were previously routed through South Africa. Furthermore, all SADC capitals are now linked directly by air. Previously such contact required connections through Johannesburg (Stoneman and Thompson, 1991).

SADCC member states also enjoyed some success in obtaining aid from international development institutions, like the World Bank, as well as various government aid agencies. In 1988, the World Bank's vice-president for Africa, Edward Jaycox, described SADCC as a 'functioning example of how regional cooperation in Africa might work'. Former UK aid minister Chris Patten commented that to 'support SADCC is to support success' (quoted in Stoneman and Thompson, 1991, p. 6). The percentage of foreign funding varied between 66 and 100 per cent for SADCC's different programmatic initiatives. SADCC's *Annual Progress Report* (1989–90) indicates that external funding of over US$3 billion had been secured for a range of projects. While SADCC's success at securing foreign funding ought to be applauded at one level, it may have displaced rather than eliminated dependency. Furthermore, the funding was related to specific foreign policy directives of OECD countries to support SADCC as a means of voicing their protest against South Africa. The problems posed by these concerns will be analysed in the next section.

PROBLEMS WITH SADCC

In spite of some of its achievements Hawkins (1992, p. 115) argues that after 10 years of existence, SADCC had made very little progress towards a reduction in external dependence, one of the major objectives outlined in its declaration. Hawkins concludes that this is hardly surprising given the necessity of the infusion of capital and technology from external sources to 'under-developed economies'. In addition, little progress was made with respect to reduction of dependency on South Africa. Thus in 1990 trade flows between South Africa and SADCC countries exceeded intra-SADCC transactions by more than 3:1, according to Hawkins (1992). While there is debate on the magnitude of trade links between South Africa and SADCC, the strategic importance of South Africa is clearly evident. Manufactured goods continue to constitute the bulk (up to 95 per cent) of imports in most SADC countries, except in Zimbabwe. South Africa supplies more than one-third of the region's imports of manufactures and commands a major share of the market with respect to spares, intermediate goods and capital equipment for the mining industry in Zambia and Zimbabwe. Furthermore, Weimer (1991) points out that intra-SADCC trade accounts for only about 5 per cent of total trade. The major impediments to improvement of this situation are the lack of tradable goods and services due to the similarity of products. In addition, serious balance-of-payment and foreign-exchange problems plague most SADC countries.

Weimar (1991) asserts that one of the major concerns with SADCC projects is reliance on foreign funding. More than 90 per cent of financing for SADCC projects came from foreign sources. This trend contradicted the two aims of the Lusaka Declaration. Donors might lose interest in supporting SADC(C) projects when a democratic regime is established in South Africa, as Simba Makoni was acutely aware (quoted by Novicki, 1990, p. 34):

> I regret to say that a good measure of our support from the international community has really not come to us on our own account. It has come to us as sympathy support against apartheid . . . So while one appreciates the amount that we have been receiving, the spirit behind the figures is not a very comforting one, because if it remains the basis for that support, when apartheid goes, then so also will the funds go with it.

Davies and Martin (1992) contend that it is highly likely that a major proportion of the aid currently funding SADC(C) projects may be diverted to support projects in South Africa. In addition, the imperative for South Africa to set up front companies in neighbouring states, in order to counteract sanctions, will be eroded in a post-apartheid era.

Finally, the difficulty of reconciling national concerns with regional ones has posed major problems for economic and political co-operation among member states. Emang Maphanyane (1990) stated that 'national planners have remained totally parochial in approach and have not provided information necessary to influence the political trade-offs that are necessary for regional integration'. Other observers have been more critical by contending that 'national chauvinism' has been the major cause of the failure on the part of member countries to integrate their economies (*Zimbabwe Press Mirror*, October 1990). Despite a shared colonial experience, there exists considerable diversity among members of SADC. This has resulted in varying degrees of commitment to SADC(C)'s agenda. Domestic élites were frequently unwilling to put the interest of the region over and above the benefits they received from national-based development projects.

While dependency on South Africa as well as South Africa's destabilization tactics have constrained the ability of southern African states to plot their own developmental agenda, the problem has been further exacerbated by the political and bureaucratic weaknesses of many SADC member states. South Africa was able to exert its influence on the region partly through the effective manipulation of already existing factions within the particular nation-states of SADCC. State legitimacy in southern Africa has been adversely affected by the absence of a unified civil society with a common definition of the public good and by the weak links between states and societies (Chazan, 1988; Rothchild and Michael, 1988). Factionalism is a widespread problem in most southern African states, which maintain their support through the careful distribution of patronage. Dunn (1986) and Sklar (1986) have argued that various strands of ethnic, religious, and regional factionalism in Africa have thwarted the development of a coherent civil society whose concerns could subsequently be addressed by the state. Regional factionalism has posed especially difficult problems for Zambia, Malawi and Lesotho (Gertzel *et al.*, 1984; Laslett, 1985). Ideological factionalism has impeded progress to varying degrees in Zimbabwe, Angola and Mozambique (Ranger, 1980; Soremekun, 1984; Hanlon, 1984). While factionalism has been less problematic in Botswana, Tanzania, and Swaziland than elsewhere, ethno-class inequalities still plague these societies (Polhemus, 1983; Crush, 1979). Strategies to contain such factionalism included the centralization of political and economic power and, in many cases, the elimination of all political opposition. This problem is aptly highlighted by Chazan (1988, p. 51):

> Centralization became a means for consolidation, but did not necessarily imply full political control. In fact, the highly concentrated system of rule created at this time was as noted for what it excluded as for what it sought to encompass . . . It was this broad pattern of distancing the state from societal constraints, of pushing the society out, that constituted the initial response of African governments to the problems of control that they faced when they took office . . .

As a consequence of the factionalism and centralization that characterize states in southern Africa, developmental efforts of the government bureaucracy were not put to the use of national interests but rather were used as a means to distribute patronage among competing factions, thus facilitating the centralization of power.

MacDonald (1990, p. 56) points out that the bureaucracies of the nine SADCC states were not uniformly weak. However problems of centralization and 'inappropriate' government intervention and structures plague all member states. In Malawi, for example, civil servants were severely constrained by the authoritarian nature of Banda's leadership. Hanlon (1986) points out that external advisers and key civil servants supplied by South Africa have seriously impeded the development of autonomous plans. Swaziland finds itself in a similar predicament, where key sectors of the economy are staffed by expatriate South Africans who have facilitated a closer linkage of the Swazi economy to that of South Africa. Furthermore, the modern–traditional sector dichotomy is maintained by effectively suppressing class discontent (Davies *et al.*, 1985).

The major conclusion that can be drawn from the preceding analysis is that while the interdependency between South Africa and SADC member states may be asymmetrical, it is not uni-directional. One of the major weaknesses that ultimately thwarted SADCC's development efforts was its inability to establish national political support for its regional development efforts. As a consequence, South Africa was able to exploit already-existing tensions and suspicions among SADCC members. There exists, contrary to SADCC's declared objectives of reduced dependence on South Africa, a complex and paradoxical network of links that ties South Africa and SADC member states together. Libby (1987) argues that the imperative for survival, irrespective of ideological commitment, is a characteristic feature of all the regimes in the region. In this respect, one needs to evaluate the political significance of economic linkages and alliances in terms of their impact on the ruling party, class, or group and the political opposition to them in each state in the region. With respect to the Malawian case for example, after surviving the cabinet crisis of 1964, Banda used economic ties with South Africa as a means to establish his personal political power base. Similar conclusions can be drawn for a number of states in southern Africa, including Zambia, Botswana, Zaïre, Malawi and, to a lesser extent, Zimbabwe and Mozambique. The earnings that SADCC countries derived from economic linkages with South Africa were (and remain) crucial for the political patronage that various southern African states need to ensure the unity of and political support for the ruling classes.

SOUTH AFRICA AND SADC: IS A COOPERATIVE FUTURE POSSIBLE?

Discussions on South Africa's future relations with other states in the region ought to be situated in terms of South Africa's regional geo-economic significance. South Africa has the strongest and most diversified economy in the region as well as the highest GDP. In 1989, its GDP was US$ 80 billion compared to the combined GDP of US$ 25 billion of the nine SADC states. South

Africa's strength is more evident when its land-mass and population size are taken into consideration. South Africa's population accounts for more than one-third of the region's total of 90.7 million. Mozambique follows South Africa with about 15 million people. Angola, Zambia, Zimbabwe and Malawi have between seven and 10 million people while Lesotho, Swaziland, Botswana and Namibia have less than two million each.

Another important contrast that can be drawn between South Africa and its neighbours pertains to the structure of the economy. South Africa's economy is well diversified with the agricultural sector representing only 6 per cent of its GDP while the remaining 94 per cent originates in the industrial (44 per cent) and service sectors (50 per cent). The national economies of southern African countries, on the other hand, show a heavy preponderance of primary production, especially mining and agriculture. Most of the economies of southern Africa are heavily dependent on one or few commodities.

In light of these strengths of South Africa, its economic advantage in the region is still considerable. South Africa historically has been a major foreign investor in the region with investments in all SADCC member states except Angola and Tanzania. In Zimbabwe for example, an estimated 25–30 per cent of privately owned capital stock is estimated to be South African, although there has been a small reduction since 1985 (SADCC, 1986). It is estimated that approximately 40 per cent of registered industrial enterprises in Botswana are owned by South Africans. In Zambia, South Africa owns key mining engineering firms and dominates the freight and forwarding business throughout the region (SADCC, 1986). The economic prosperity of South Africa's neighbours is dependent, to a significant degree, on the political and economic posture of a future democratic South Africa towards regional co-operation and co-ordination. The recent changes in South Africa offer possibilities for increased co-operation between South Africa and its neighbours to the north. Recently, there has been much debate on when and under what circumstances South Africa ought to be admitted into SADC and what are the possible post-apartheid peace dividends for the region. The goal of 'reducing dependence on South Africa' was omitted in SADC's new policy prescriptions in the August 1992 Summit meeting. The Summit also advanced broad proposals to move away from project co-operation to close political co-operation in order to facilitate the establishment of equitable trade integration. It was envisaged that this would involve the institutionalization of policy measures that reduce tariff barriers for intra-SADCC trade, promote a less-restricted movement of capital and people, as well as the creation of regional infra-structural authorities and possibly a development bank (Maasdorp, 1992; SADCC, 1992).

The African National Congress (ANC) has not yet advanced any concrete proposals on future relations with southern African states. However, various policy discussion papers and statements by senior ANC officials indicated that an ANC-ruled South Africa would seek membership in SADC and foster a co-operative relationship with other states in the region. In a policy discussion paper the ANC has stated that (ANC, 1991, pp. 15–16)

A future democratic government should actively seek to promote greater regional cooperation along new lines which would not be exploitative and which will correct imbalances in current relationships. The new state must be prepared to enter negotiations with its neighbours to promote a dynamic and mutually beneficial form of cooperation and development. While all of us stand to benefit from such an arrangement, it should be recognized that creating a new non-exploitative form of regional cooperation will require prioritising the interests of the most impoverished of our neighbours in certain areas, according to principles of affirmative action.

These statements do not state explicitly the policies that will be pursued by an ANC-led government in South Africa. Initial steps, according to Davies (1989) could possibly include assuring the rights and wages of foreign workers and willingness to assist in the rebuilding of neighbouring transportation systems. The end of apartheid may afford new trading opportunities in the region. South Africa, in all likelihood, will purchase Angolan oil. Regional co-operative efforts with respect to transportation, agricultural research, as well as electrical and water supplies, are possible. Beyond the possibilities of these co-operative measures, however, a more comprehensive regional development strategy will have to confront the uneven tendencies of capitalist development as well as the political tensions and contradictions that characterize the region.

Behind the political rhetoric of new conditions for co-operation between South Africa and other southern African states in a post-apartheid era, there are underlying tensions with respect to the role South Africa will then play in the region. The economic imbalance between South Africa and its neighbours poses problems for translating good intentions into actual policy. There is a growing fear that South Africa, however governed, will exercise considerable economic leverage over its neighbours. On his first American tour, Nelson Mandela participated in discussions with the Rockefeller Foundation and American corporate executives about the establishment of a new South African Development Bank after majority rule has been achieved (Kifner, 1990). Weisfelder (1990, p. 14) contends these unilateral initiatives on the part of the ANC have raised concern over its commitment to regional versus nationally based development initiatives. Weisfelder asks the important question of whether post-apartheid South African involvement in SADC 'will be an important foreign policy commitment, or just a *pro forma* exercise'. Weisfelder goes on to argue that SADC receives very little news coverage in South Africa, even in the progressive press and media. For instance, the *Weekly Mail* (Johannesburg) produced only one short article to commemorate the 10th anniversary of SADCC. Furthermore, the majority of the rank and file of South Africa's population is unaware of how South Africa might benefit from regional co-operation. Thus it is likely that a new government in South Africa will face an electorate that is concerned more with domestic bread and butter issues than the allocation of scarce resources to regional co-ordination efforts.

Critical readings of statements by SADC(C) officials and analysts on South Africa's future role reflect an ambivalence that is related to the fear of South Africa's potential domination in SADC. Simba Makoni asserts that:

> There is indeed a role for a free and democratic South Africa in SADCC *as an equal partner, not an overlord and domineer*; to contribute positively to our joint endeavours toward a balanced and equitable regional integration. A free and democratic South Africa, *accepting and respecting the norms and mores of international and regional partnership and interdependence*, has a guaranteed place in the SADCC family (quoted in *Southern African Economist*, 1990, p. 23).

Colleen Morna (1990b) cites a Zimbabwe economist who argues that the entry of South Africa could diminish the roles played by other states in SADCC:

> When South Africa comes in, not only will the size of the pond increase dramatically, but it will be as though a whale—maybe even a shark—has been put into it. Although this is not a popular thing to say right now, it is not clear that the economic impact of a free South Africa will be less harmful than the economic impact of overt destabilisation.

The August 1992 SADC Summit Meeting stressed that regional co-operation in southern Africa should be based on the principles of 'equity' and 'mutual benefit' (SADC, 1992, p. 5). This can be interpreted as a realization among SADC members that the economic legacies of apartheid for the region are likely to persist into the foreseeable future and that the regional policies of an ANC-led government may not deviate substantially (except for overt destabilization) from those of the National Party.

The internationalization of production and the new patterns of regional development are contributing to new patterns of international migration, especially between peripheral regions and core centres. In sum, the world's more developed areas (the wealthier states of Western Europe, the United States and, to a lesser extent, core countries of the Pacific realm) are experiencing a greater influx of illegal migrants from declining regions and peripheralized countries (Sassen, 1988). The issue of migrant labour from southern African states into a post-apartheid South Africa will probably be one of the most difficult to address in any attempt to structure a mutually beneficial regional order in southern Africa (Davies, 1990). In 1992, for example, some 500 000 citizens of independent states were legally registered as migrant workers in South Africa. For Lesotho, Botswana, Mozambique, Swaziland and Malawi, migrant workers in South Africa make up a sizeable proportion of the total wage-earning population (SAIRR, 1992). Furthermore, foreign-exchange earnings from migrant labour are an important source of income for certain SADC member states. In addition, it is estimated that there are some 1.3 million 'illegal aliens' from SADC member states working in South Africa. A new democratic regime in South Africa will probably have to confront an unprecedentedly large influx of refugees and migrants into a post-apartheid South Africa. There is the possibility that large-scale movements of populations can take place across national boundaries in southern Africa. Ashforth (1992, p. 370) argues that most of the present populations of Lesotho, Botswana, Swaziland, Zimbabwe and Zambia have the necessary language skills to seek work in urban South Africa. However, an influx of these groups to South African urban areas could potentially exacerbate the ethnic tensions in South African cities. A post-apartheid government may be under severe domestic pressure to limit foreign labour migration into South Africa and, as a consequence of national self-interest, exclude other southern Africans from the centres of economic growth in South Africa. Furthermore, cross-border investment may decline in the immediate post-apartheid era since the motive for setting up front companies in Botswana and Swaziland may be offset by the ending of sanctions (Hawkins, 1992).

Thus, the establishment of a democratic regime in South Africa may not in itself guarantee political stability and conditions for economic growth. Different possible combinations of the aforementioned conditions may impact South Africa's ability and willingness to participate in economic and political alliances within the region. Certain analysts and political figures seem to hold on to the rather simplistic view that in a post-apartheid era, the South African economy will experience rapid growth that is bound to have positive impacts on the region as well. However, the economic prospects of a future South Africa are tenuous at present. Real growth has declined from more than 5 per cent per annum in the sixties to less than 2 per cent in the eighties and nineties. The South African economy produced eight jobs for every hundred people entering the labour market in the second half of 1992, a trend which is unlikely to improve considerably in the near future (*The Economist*, March 1993). A future government in South Africa may develop, out of concern for domestic issues and problems, more inward, nationally based policies. Regional co-operation may be premised either on narrowly perceived national interest or tokenism rather than genuine concern for the other states in the region.

TOWARD A REGIONAL DEVELOPMENT STRATEGY

The preceding discussion illustrated that in spite of SADC(C)'s success in the spheres of transport integration and communications, the larger goals of dependency reduction and economic integration were complicated by the intricacies of the southern African regional political economy. After a series of economic and political crises in the domestic sphere and transformations in the regional and international environment, SADC states have adopted new economic strategies to deal with the rapidly changing global and regional political economy. Under the impetus of the IMF and World Bank structural adjustment programmes, all SADC members have adopted economic liberalization programmes and are trying to reduce the involvement of the public sector in development initiatives. Furthermore, the demise of the Soviet Union and the popular as well as institutional repudiation of radical Maoism in China have led to a re-evaluation of the more utopian development agendas that have characterized certain southern African states. The changes in the global political scene, especially the end of the Cold War, have significantly altered the geostrategic climate in the region. International interest in Africa is generally on the decline. Sub-Saharan Africa is being increasingly marginalized from the dynamics of the international economy. Transnational corporations that would have invested in southern Africa see new opportunities in Eastern Europe; many Eastern European countries have similar economies to those in southern Africa. One important sign of this marginalization is the recent trend of British investment in the region. Coker (1991, p. 117) argues that 'there has been a sizable and possibly irreversible disengagement from southern Africa by British companies'. Zimbabwe, for example, accounts for 25 per cent of British industrial investment in Africa. However, some 37 British companies have withdrawn since 1985. More than 17 companies have disinvested from Zambia. Sub-Saharan Africa now accounts for only 0.5 per cent of British foreign industrial investment. This is in contrast to nearly 5 per cent 10 years before. Furthermore, southern African economies may have to compete for aid in an expanded international environment.

In the light of South Africa's more immediate developmental concerns, it is debatable whether a non-racial, democratic South Africa will undertake regional developmental initiatives given its own internal crises and contradictions. On the other hand, a post-apartheid democratic regime may want to ameliorate the tensions and schisms that historically existed between South Africa and its neighbours during the apartheid era. Taking into consideration the logistical and other support of southern African states for the major liberation movements in South Africa, an ANC-led government cannot afford to antagonize neighbouring countries by following policies similar to those pursued during the apartheid era. Sethi (1991) argues that if southern Africa is to deal effectively with its poor economic performance, some form of regional alliance is indispensable. In the absence of such an attempt, existing regional hierarchies and inequalities are likely to prevail. Some of the important mechanisms for effective regional alliances, according to Sethi, include a large and integrated consumer market. Although this larger market offers tremendous growth potential and consumer purchasing power, the mechanisms for effective transnational competition and co-operation ought to be developed. There is urgent need for the rapid expansion of physical infrastructure such as railroads, highways, telecommunications and air-traffic. It may be necessary to expand and restructure existing regional co-ordination agencies to develop and manage infrastructure projects. A number of models in the United States and elsewhere offer good possibilities for adaptation and replication. The New York–New Jersey Bridge and Tunnel Authority, which is jointly controlled by the states of New York and New Jersey, is one such example. Scandinavian Airlines Systems (SAS) provides an example of multi-country co-operation in owning airlines with both national and international networks. Within SADCC itself, the Beira Corridor project is another example of how regional co-ordination can be expanded for the benefit of the entire region. Perhaps the feasibility of regionally dispersed and co-operatively managed high value-added industries should be investigated. There exists the need for southern Africa to formulate a viable urban and regional development strategy that transcends national boundaries in the region, if it is to deal effectively with the historical legacies of uneven development. In the absence of a regional development policy, industries are most likely to be concentrated in areas with readily available infrastructure and skilled human resources. Such an event may benefit South Africa, at the expense of other less-developed states in the region. Furthermore, without effective regional alliances, each country will probably develop its own small-scale technological project which may be little more than a status symbol. These projects may subsequently drain these countries' meagre economic resources rather than create momentum for economic and technological growth.

Afriyie (1991, p. 11ff) argues that a regionally dispersed production system that transcends existing political boundaries will be the most effective method for regional co-operation and co-ordination on the African continent. Such a system, according to Afriyie, will generate greater employment opportunities across a larger economic space and simultaneously exploit factor endowment differences in the various regions. A dispersed production system will also reduce the political tensions caused by firms locating in one particular region. Afriyie acknowledges, however, that there are considerable hurdles to be overcome before the realization of dispersed production systems. While some of the constraints are rooted in economic factors, the major constraint to regional economic integration, according to Afriyie, lies in the lack of political will to implement such policies. Gourevitch (1986) perceptively notes that 'economic theory can tell us a lot about policy alternatives, but unless our economics contains an understanding of power, it will not be enough to understand the policy choices actually made'. While there is a justifiable need for greater regional co-operation to reduce inequalities, the *realpolitik* of the region may necessitate prioritizing national interest over regional concerns.

At a more fundamental level discussions on regional integration and economic growth ought to be situated in the context of how economic development policies (regional or national) assist or impede democratic participation in the political process. Barber (1993, p. 12) appropriately argues that 'in the late twentieth century, the world is not in a position to privilege [sic] economic growth over political democratization . . . we have to find ways to create a trade-off between values of the market and the demands of political, legal and civil rights.' One of the paradoxes of the present political restructuring in southern Africa is that South Africa is likely to have more democratic institutions in place prior to many of its southern African and African counterparts. Many Third World countries responded to the United Nations' Human Freedom Index by advocating a ban on the publication. Poorer nations, classified as the least free, were apprehensive of how the Index might be used by international development agencies to link development assistance to a country's human rights record (Lewis, 1992, p.A6; United Nations, 1991). Faced with the prospect of the increasing linkage of aid to democratic reform, states in the region will be hard pressed to adopt political and economic reforms, perhaps not in conditions of their own choosing. Democratization alone, however, will not address the fundamental issues of uneven development that characterize the Third World. In Nicaragua, for example, there is increasing dissatisfaction with President Violeta Barrios Chamorro among her peasant allies. This discontent, particularly over the government's failure to fulfil pledges of land redistribution, has led to a rising spiral of violence and retaliation that could drive thousands of former rebels to take up arms, just months after their demobilization. In Eastern Europe, unrealistic expectations that the fall of communist regimes would instantly improve living standards, are generating new forms of conflict and instability. Extreme nationalism and religious fanaticism seem to be filling the void left by the collapse of communism.

With particular reference to the southern African context, the recent reforms in South Africa have already begun to place severe tensions on the fragile cohesion among SADC members. The Zambian government, for example, is eager to exploit ties with South Africa as soon as possible. Chiluba, the current president of Zambia, stated during a recent PTA (Preferential Trade Area) meeting that 'there is no reason why the South African government should not be invited to attend PTA meetings as observers . . . we cannot close our eyes to the facts. Increasing contact with South Africa is a fact of life.' Other members expressed dismay in forging direct links with the white regime in South Africa. These tensions are indicative of the attempts by various political actors to form new alliances in the region. While some of these alliances are likely to benefit the region, there exists the possibility that reactionary elements, whose power bases have been eroded due to democratic transition in the region, could constitute themselves into a small yet destructive hegemonic influence. South African military leaders could form working alliances with individuals like Jonas Savimbi or groups like the Mozambique National Resistance (MNR), thereby causing new forms of destabilization in the region. While the numerous developmental problems that characterize southern Africa require regionally-based solutions, it remains to be seen whether political leaders in southern Africa have the commitment to transcend the confines of nationalism and work for the common good. In a recent book on African identities and cultures, Appiah (1992) asserts that 'we are all contaminated by each other in a complex, interdependent world that is ill-served by the dead-end effort of engaging in the manufacture of otherness'. While Appiah's philosophical argument offers valuable insight that cannot be ignored in future development strategies, the manufacture of 'otherness' historically has been a politically expedient means of maintaining power and privilege. Current trends suggest that reconstituted hegemonic hierarchies are likely to prevail in the political economy of southern Africa for the near future. In this respect, it seems Appiah's vision of a universal moral framework for addressing socio-political issues is likely to prove elusive in an era of contingency, fragmentation and retribalization in developing societies.

ACKNOWLEDGEMENTS

I would like to thank Kavitha Ramachandran and Amin Khadr for their comments on an earlier draft of this paper. The Dean's Office of Rutgers College and the Department of African Studies provided financial assistance which enabled me to attend the Second Southern African Geographers' Symposium at Oxford where this paper was initially presented.

REFERENCES

Afriyie, K. (1991) 'The African Economic Crisis: A Regional Approach to Capacity-Building via Development of Technological Infrastructure', paper presented at the annual meeting of the African Studies Association, St Louis, Missouri.

ANC (1991) *Discussion Document: Economic Policy*, Department of Economic Policy, Johannesburg.

Appiah, K.A. (1992), *In My Father's House: Africa in the Philosophy of Culture*, Oxford University Press, New York.

Ashforth, A. (1992) 'South Africa: reconstructing an imperial state', *Dissent*, 39(3), Summer, 370–7.

Barber, B. (1993) 'Democracy and civic education for the twenty-first century', *Common Purposes: the journal of the committee to Advance our Common Purposes (Rutgers University, USA)*, 4(1), 12–15.

Blumenfeld, J. (1991) *Economic Interdependence in Southern Africa: From Conflict to Cooperation*, St. Martin's Press, New York.

Chazan, N. (1988) 'State and society in Africa: images and challenges', in D. Rothchild and N. Chazan (eds), *Precarious Balance: State and Society in Africa*, Westview Press, Boulder, Colorado.

Coker, C. (1991) 'What future for Southern Africa?', *The World Today*, 47(7), July, 116–20.

Crush, J. (1979) 'Parameters of dependence in southern Africa: a case study of Swaziland', *Journal of Southern African Studies*, 4(1), 55–66.

Davies, R. (1989) 'Some implications of competing post-apartheid scenarios for the Southern African region', paper presented at a seminar sponsored by Fernand Braudel Center, SUNY, Binghamton, New York.

Davies, R. (1990) 'Post-apartheid scenarios for the Southern African region', *Transformation*, 11, 12–39.

Davies, R. and Martin, G.M. (1992) 'Regional prospects and projects: what futures for Southern Africa', in S. Vieira, W.G. Martin and I. Wallerstein (eds), *How Fast the Wind? Southern Africa, 1975–2000*, Africa World Press, Trenton, New Jersey.

Davies, R.H., O'Meara, D. and Dlamini, S. (1985) *The Kingdom of Swaziland: a Profile*, Zed Books, Totowa, New Jersey.

Dunn, J. (1986) 'The politics of representation and good government in post colonial Africa', in P. Chabal (ed.), *Political Domination in Africa*, Cambridge University Press, Cambridge.

Evans, P. and Stephens, J.D. (1988) 'Studying development since the sixties: the emergence of a new comparative political economy', *Theory and Society*, 17(5), 713–45.

Evans, P.B., Rueschemeyer, D. and Skocpol, T. (eds) (1985) *Bringing the State Back In*, Cambridge University Press, Cambridge.

Gertzel, C., Baylies, C., Szeftel, M., (eds), (1984), *The dynamics of the one party state in Zambia* (Introduction), Manchester University Press, Manchester.

Gourevitch, P. (1986) *Politics in Hard Times: Comparative Responses to International Economic Crises*, Cornell University Press, Ithaca.

Haarlov, J. (1988) *Regional Cooperation in Southern Africa: Central Elements of the SADCC Venture*, African Studies Institute, Copenhagen.

Haggard, S. (1990) *Pathways from the Periphery*, Cornell University Press, Ithaca.

Hanlon, J. (1984) *Mozambique: the Revolution under Fire*, Zed Books, London.

Hanlon, J. (1986) *Beggar your Neighbours: Apartheid Power in Southern Africa*, CIIR, London.

Hawkins, A.M. (1992) 'Economic development in SADCC countries' in G. Maasdorp and A. Whiteside (eds), *Towards a Post-apartheid Future: Political and Economic Relations in Southern Africa*, St Martin's Press, New York.

Jessop, B. (1990) *State Theory: Putting Capitalist States in their Place*, Pennsylvania State University Press, University Park, Pennsylvania.

Kifner, J. (1990) 'Planning is begun on South Africa aid bank', *New York Times*, 5 July.

Laslett, R. (1985) *An Account of Malawi's Economy and the Economic Policy in the 1970s*, Proceedings for a seminar held at the Centre for African Studies, University of Edinburgh, No. 25.

Lewis, P. (1992) 'UN freedom index angers third world', *The New York Times*, 23 June, A6.

Libby, R.T. (1987) *The Politics of Economic Power in Southern Africa*, Princeton University Press, Princeton.

Lipietz, A. (1987) *Mirages and Miracles*, Verso Press, London.

Maasdorp, G. (1992) *Economic co-operation in Southern Africa: Prospects for Regional Integration*, The Research Institute for the Study of Conflict and Terrorism, London.

MacDonald, H. (1990) *Planning Reconstruction and Reform on a Regional Scale: the Efforts of the Southern African Development Coordination Conference*, unpublished Ph.D. dissertation, Rutgers University, New Brunswick, New Jersey.

Maphanyane, E. (1990) 'Economic Development in Southern Africa—SADCC Perspective', paper presented at the conference on Rethinking Strategies for Mozambique and Southern Africa, Maputo, April, 21–24.

Meyns, P. (1984) 'The Southern African Development Coordination Conference SADCC and regional cooperation in Southern Africa' in D. Mazzeo (ed.), *African Regional Organisations*, Cambridge University Press, Cambridge.

Milkman, R. (1979) 'Contradictions of semi-peripheral development: the South African case', in W.L. Goldfrank (ed.), *The World System of Capitalism: Past and Present*, Sage Publications, Beverly Hills.

Morna, C.L. (1990a) 'SADCC's first decade', *Africa Report*, May–June, 49–52.

Morna, C.L. (1990b), 'Zimbabwe braces for shifts in regional leadership role', *Christian Science Monitor*, 12 July.

Novicki, M.A. (1990) 'A decade of regional cooperation', *Africa World Report*, July–August, 34–9.

Polhemus, J.H. (1983), 'Botswana votes: parties and elections in African democracy', *Journal of Modern African Studies*, 21(3), 397–430.

Ranger, T. (1980) 'The changing of the old guard: Robert Mugabe and the revival of ZANU', *Journal of Southern African Studies*, 7(1), 71–90.

Rothchild, D. and Michael, W.F. (1988) 'African states and the politics of the inclusive' in D. Rothchild and N. Chazan (eds), *Precarious Balance: State and Society in Africa*, Westview Press, Boulder, Colorado.

SADCC (1986) *Macro-economic Survey*, Gaborone, Botswana.

SADCC (1992) 'Declaration towards the Southern African Development Community', Gaborone, Botswana.

SAIRR (1992) *Race Relations Survey 1991–2*, South African Institute of Race Relations, Johannesburg.

Sassen, S. (1988) *The Mobility of Labour and Capital: a Study of International Investment and Labour Flow*, Cambridge University Press, Cambridge.

Sethi, S.P. (1991) 'Prospects for two-way exchange in the economic arena between post-apartheid South Africa and the rest of Africa', paper presented at the Conference on the Challenges of Post-Apartheid to Southern Africa in Particular and Africa in General, sponsored by the Africa Leadership Forum, Windhoek, Namibia, 8–10 September.

Shafer, M. (1994) *Sectors, States and Social Forces: towards a New Comparative Political Economy of Development*, Cornell University Press, Ithaca, forthcoming.

Sklar, R. (1986) 'Democracy in Africa' in P. Chabal (ed.), *Political Domination in Africa*, Cambridge University Press, Cambridge.

Soremekun, F. (ed.) (1984) *The Political Philosophy of African Foreign Policy*, Gower, Aldershot.

Stoneman, D. and Thompson, C.B. (1991) *Southern Africa after Apartheid: Economic Repercussions of a Free South Africa*, Africa Recovery Briefing Paper No. 4, United Nations Department of Public Information, New York.

Thompson, C. (1986) 'SADCC struggle for economic liberation', *Africa Today*, July–August, 59–63.

Tostensen, A. (1993) 'What role for SADC(C) in the post-apartheid era?' in B. Oden (ed.), *Southern Africa after Apartheid*, The Scandinavian Institute of African Studies, Uppsala.

United Nations (1991) *Human Development Report*, Oxford University Press, New York.

Weimer, B. (1991) 'The Southern African Development Co-ordination Conference SADCC: past and future', *Africa Insight*, 21(2), 78–88.

Weisfelder, R.F. (1990) 'Collective foreign policy decision-making within SADCC: do regional objectives alter national policies?', paper presented at the 33rd Annual Meeting of the African Studies Association, Baltimore, Maryland, 1–4 November.

11 The Relevance of the European Approach to Regional Economic Integration in Post-apartheid Southern Africa

RICHARD GIBB

The collapse of statutory apartheid and the removal of South Africa's pariah status will have a number of profound implications for sub-Saharan Africa as a whole and for southern African integration in particular. Despite the momentous transformation taking place in southern Africa, and the potential significance of the political, social and economic reforms occurring throughout the region, relatively little attention has been devoted towards examining the nature and evolution of regional economic integration in the 'new' southern Africa (Tjonneland, 1989; Davies, 1991). Notwithstanding Africa's dismal track record in the field of regional economic integration, well illustrated by the collapse of the East African Community and the failure of the West African Economic Community (WAEC) and the Economic Community of West African States (ECOWAS) to liberalize and promote intra-regional trade (Davenport, 1992), most of the governments within southern Africa support the idea of closer economic co-operation as a means of promoting economic development and reducing the exploitative dependency relationships arising from limited and unspecialized internal markets which lack economies of scale (Gibb, 1993). Regional economic integration is perceived to be an appropriate mechanism for the expansion of trade, income and bargaining power.

Since the early 1960s, South Africa's foreign policy has been concerned with promoting further regional economic integration throughout the sub-continent. The 'outward policy', *détente* and CONSAS[1] all strove to establish diplomatic links and enhance economic ties with the African continent and southern Africa in particular. By establishing a 'co-prosperity sphere', a common market or a 'constellation of independent states', South Africa tried to promote regionalism, albeit centred upon Pretoria, in the firm belief that economic factors would help remove regional conflicts and tensions associated with apartheid. However, for the majority-ruled states of the region, South Africa's tactics amounted to little more than 'apartheid as a foreign policy' (Green, 1981). Up to 1989–90, attempts to evaluate the relevance of transplanting European models of regional integration to southern Africa were therefore both fundamentally and structurally

The Geography of Change in South Africa, edited by Anthony Lemon.
© 1995 by the Editor and Contributors. Published in 1995 by John Wiley & Sons Ltd.

inappropriate (Scheepers, 1979; Vale, 1982). This led Holland (1983) to suggest that parallels with the European Community were 'deceitful'. The conclusions reached by Holland warrant being quoted at length:

> Despite the superficial similarities between the customs unions operated in Europe and southern Africa, the comparison is fraudulent . . . the simple political reality of apartheid, more than anything else, exemplifies the ludicrous incongruity of the comparison. Recognition, let alone membership of the Constellation of states by Black Africa, would be regarded as an act of obsequious acquiescence tantamount to condoning apartheid . . . Conversely, a constellation of states based on the Southern African Development Coordination Conference countries could legitimately draw from the Community's experience: but such a prospect is so remote as to be a chimera' (Holland, 1983, p. 46).

However, this 'chimera' is in the early 1990s almost reality and provides the rationale underpinning the current chapter. The aim of this study is not to promote a Eurocentric approach to regional economic integration in southern Africa. The principal objective of the chapter is to highlight ways in which the region could benefit from the European experience of economic integration. The chapter does not therefore examine the rationale supporting regional economic integration (see for example: Gibb, 1993; Langhammer, 1992; Matthews, 1984). Focus is concentrated upon the theories, and associated mechanisms, of integration.

Before proceeding to evaluate critically the relevance of the European model, it is first worth emphasizing that regional economic integration is as much a political process as it is an economic one. Inevitably, economic integration and political sovereignty are issues that are inextricably interlinked and at the heart of any regional initiative, in Europe or elsewhere. The temptation to examine regional economic integration as a purely technical problem must therefore be avoided. However, integration is often pursued through detailed and technical economic legislation. For example, the word 'political' was never used in the Treaty of Rome that established the European Economic Community on 25 March 1957. However, signatories to that Treaty were under no illusion as to its essentially political character. More recently, Lord Cockfield's White Paper on 'Completing the Internal Market', that led to the Single European Act (SEA) of 1987, was a technical and pragmatic document focusing on a specific, limited aim. It was originally perceived by many to be a rather technical issue over which the member states of the Community could be persuaded to co-operate in order to ensure a genuine common market within which economies of scale could be achieved and the economic challenge of Europe's competitors met (Cecchini, 1988). However, the 'technical issue' of removing non-tariff barriers increased pressure for an overall harmonization of Community law such as in social policy, the environment, health and welfare. By requiring the elimination of economic barriers, the Single Market programme inevitably unleashed forces to merge national jurisdictions into common European regimes.

The economic and political components of integration are therefore, on the one hand, so interlinked as to make any distinction between the two meaningless. On the other hand, it is often easier for states to pursue policies of economic integration and manage the delicate issue of political sovereignty as if it were a technical economic matter. These two perceptions of integration are by no means contradictory and can be applied to both the federalist and functionalist approaches to integration examined in the present chapter.

THE EUROPEAN COMMUNITY'S APPROACH TO INTEGRATION

It is neither possible nor desirable to review the voluminous literature devoted to the theoretical and practical approaches to European economic and political integration. Examining the prospects for transplanting European models of regional integration to Southern Africa, Vale (1982, p. 31) observes that: 'theoretical interest in integration in international relations has been strong. Approaches and theories to an understanding and explanation of the process abound—almost as many theories as theorists.'

Although European union is largely perceived to be a post-1945 phenomenon, the idea of integrating Europe has been a persistent theme throughout Europe's history. As early as the 14th century, Pierre Dubois proposed a 'European federation' based on Christian principles and governed by a European Council (Perry, 1984). European history has since been littered with such proposals. However, most of the contemporary theories are based upon interpretations of the 'federalist' and 'functionalist' intellectual traditions concerned with European integration in the immediate aftermath of the Second World War.

Federalism has a long intellectual history in Europe but received fresh impetus following the destructive impacts of the two World Wars. In 1946 at Zurich, Churchill called for a kind of 'United States of Europe' (Perry, 1984). Although this call was misinterpreted, as the United Kingdom remained fundamentally opposed to the sharing of sovereignty and eager to distance itself from attempts to promote formal integration, it helped stimulate a renewed interest in some form of regional co-operation along federalist lines. Consequently, a European Union of Federalists (EUF) was formed in 1946. A basic strategy of its adherents was to establish federal institutions throughout Europe to weaken the continent's nation states and help create a new European order (Wise and Gibb, 1992). Federal institutions would restrict, and perhaps even eliminate, nation state sovereignty, thus precipitating a shift of popular loyalties from national to a federal European level. Federalism, although promoting common political institutions to create co-operative unity among diverse groups, does not seek to eliminate diversity or the political institutions associated with it.

Confederalism, on the other hand, while having similar aims to federalism stresses the desirability of paramount powers resting with the political institutions of constituent member states. However, the term 'federalists' in the European arena incorporates those groups who want substantial powers for common central institutions (the 'true' federalists) and those who favour very loose decentralized confederal arrangements (Wise, 1994). In 1948 a collection of political leaders and pro-European activists met at the Hague conference to pursue the federalist cause. However, the idea of ceding national sovereignty in one mighty constitutional stroke proved impractical. The Hague conference was a severe disappointment for the European federalists. It helped establish the Council of Europe, an intergovernmental organization of very limited scope that still exists today. Although the Council of Europe is still adhered to by all West European states, it remains almost totally unknown to their citizens (Wise, 1994). Its decisions are taken by unanimous and not majority voting and its powers are mainly consultative. The Council of Europe proved to be an exceptionally weak instrument for promoting regional economic integration. Mitrany, one of the major proponents of the rival functionalist approach to integration, described federalists thus:

> Federation was an advance on empire as a way of joining under a common government a group of separate territorial units. But federation is not only inadequate but irrelevant when the general task is not to consolidate but to loosen the hold of the territorially sovereign conception of political relations (Mitrany, 1965, p. 129).

Clearly functionalists consider the direct federalist strategy to be politically unrealistic and/or undesirable. The term 'functionalism', in the European context, owes its origin to the considerable writings of David Mitrany who has been working on this topic since the early 1930s (Mitrany, 1933, 1966, 1975). The objective of the functionalists is similar to those of a federalist persuasion: to establish a new world order by reducing the sovereign powers of nation-states. However, in sharp contrast to the federalists, Mitrany advocated the concept of authority being linked to a specific functional activity. For Mitrany, the federalist approach to integration would result in a world dominated by superstates likely to reproduce national rivalries but on a larger scale.

The functionalist model offered a more pragmatic approach to integrating sovereign states within structures where conflicts could be peacefully resolved. Functionalism explicitly avoids conflicts with nationalism. As Mitrany observes:

> . . . The functional approach does not offend against the sentiment of nationality or the pride of sovereignty. Instead of a fictitious formal equality it offers even to the weakest of countries the assurance of non-discrimination and of an equality of opportunity in the working benefits of any functional activity in which it participates . . . because functional arrangements have the patent virtue of technical self-determination, the range of their task can be clearly defined and this, in turn, makes plain the powers and resources needed for its performance (Mitrany, 1965, p. 139).

Functionalism therefore advocates the establishment of a network of overlapping agencies, each with a specific task or 'function'. The territory or spatial coverage of these agencies would vary according to functional role. By the voluntary surrender of more and more sovereignty to these agencies, states would eventually find themselves inextricably interwoven into an international environment which would imperceptibly erode national sovereignty.

Although the federalist and functionalist approaches to integration are often interpreted as distinct and separate, the founders of the first European Community in 1950, the European Coal and Steel Community (ECSC), tended to merge elements of both. Believing that European unity could not be achieved in one mighty stroke by the imposition of some great federal constitution, Jean Monnet and Robert Schumann, the founding architects of the Community, pursued a 'step-by-step' approach that would erode sovereignty but at the same time satisfy national self-interest. This 'step-by-step' approach to integration is closely associated with 'neo-functionalism' and forms part of the important 'spillover' strategy (Wise and Gibb, 1992).

The spillover theory has permeated much academic literature on the process of European unity (Lindberg, 1963; George, 1985; Iconescu, 1972). Put simply, spillover refers to the process where action in one area of political/economic activity, for example coal and steel, builds up pressures to integrate in other areas, for example transport and working conditions. The Community's history is thick with examples illustrating the spillover concept. Policies designed to pursue the seemingly simple objective of creating a free-trade area create pressures that necessitate further legislation drawing states inexorably into ever closer political and economic union. Thus the Community's drive to create a single internal market by 31 December 1992 unleashed a whole range of forces that aimed to merge national jurisdictions into common European regimes, illustrating well the spillover workings of the neo-functionalist integration process (Wise and Gibb, 1992).

However, the pertinent question that needs to be answered is the extent to which the processes and predictions of the spillover theory of integration are borne out in reality. Commenting on integration literature more generally, Vale notes that:

> . . . while writing on the topic has been voluminous, its practical impact on the integration process in Europe itself has been minimal . . . The political practitioner has always . . . to balance a commitment to the common [integration] endeavour against the need to account to his [sic] constituency . . . The theoretician . . . in an eagerness to foster the 'ideal' of integration has not been able to match . . . the necessary caution of the practitioner. Thus the tension is obvious (Vale, 1982, p. 33).

However, in a recent study of the spillover forces emanating from the 1992 programme, Wise and Gibb refute this observation as far as the spillover theory is concerned, noting that:

The spillover theory could be dismissed as excessively academic. However, practical people concerned with the changing of the political shape of Europe have subscribed to its tenets, whether knowingly or not. Most notably Jean Monnet and Robert Schumann, the founding architects of today's Community, thought in terms very akin to those sketched out above [spillover]. They may have used the French term 'engrenage' (literally an enmeshing of wheels) to describe a progress where European unity would be built up progressively step-by-step . . . but the intellectual concept is essentially the same. (Wise and Gibb, 1992, p. 34).

The spillover theory therefore identifies a simple but forceful logic based upon a step-by-step approach which satisfies both national self-interest and the need to pool sovereignty. The neo-functionalist approach is well illustrated by the ECSC which, while being concerned in the short term with the intermeshing of the coal and steel industries, was unambiguously aimed at the creation of a 'European federation' in the long term.

The European experience of regional integration highlights the advantages of the neo-functionalist approach supported by spillover forces. As the founding architects of the Community recognized, it is far easier to promote integration through economic legislation and manage the sovereignty issue as an appendage to that legislation. The success of the SEA, designed to remove all impediments to free trade, contrasts sharply with the difficulties encountered by the Maastricht Treaty on European Union. Both pieces of legislation involve a considerable loss of national sovereignty for member-state governments. Indeed, the SEA transferred more power from national governments to European institutions than is envisaged by the Maastricht Treaty. However, because the SEA was perceived to be an economic and technical matter it encountered little in the way of resistance from national governments.

The present chapter now goes on to examine the approach adopted by the Southern African Development Community (SADC) towards regional integration, and then evaluates that approach in the light of the European experience.

THE SOUTHERN AFRICAN DEVELOPMENT COMMUNITY

Southern Africa has three major regional groupings: SADC, the Southern African Customs Union (SACU) and the Preferential Trade Area (PTA). The Southern African Development Coordination Conference (SADCC) was replaced by SADC in August 1992. At present, eight SADC countries are members of the PTA. Four SADC states are members of SACU, while three of these are also members of the Rand Monetary Area. The overall picture is therefore a confusing one (Figure 11.1), with many states belonging to organizations that have different, and often conflicting, objectives (Blumenfeld, 1991). For example, SACU exists to promote economic integration between South Africa and its neighbouring states while in the 1980s SADCC's primary objective was the converse: to reduce trade and

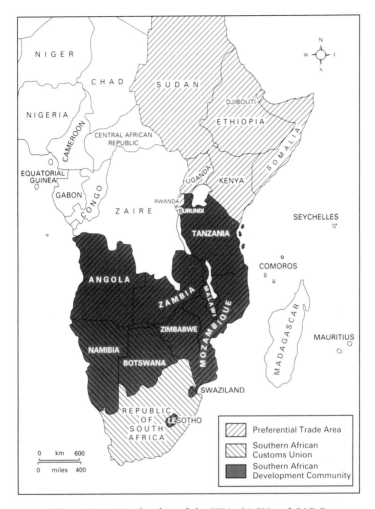

Figure 11.1. Membership of the PTA, SACU and SADC

dependence on South Africa (SADCC, 1981) through a command-type control of the economy. The new SADC aims to promote regional economic co-operation and integration on 'the basis of balance, equity and mutual benefit' (SADC, 1992a). The PTA, on the other hand, aims to promote intra-regional trade via trade liberalization and free-market principles. The present chapter does not attempt to provide an analysis of the success or otherwise of these organizations (see Chapter 10 du Pisani, 1991; Maasdorp, 1986, 1989, 1992; Davies, 1990a, 1990b).

This chapter focuses exclusively on the nature and evolution of SADC as the organization most likely to promote regional economic integration in the 'new' southern Africa (including South Africa). This is a contentious statement and disputed by many regional scholars. For example, Maasdorp (1989) regards an expanded and reorganized SACU as offering the most appropriate structure for further cooperation. Leistner (1992) on the other hand promotes the OECD as a model framework for a southern African community. It is not the purpose of this study to examine in detail the pros and cons of the existing regional organizations nor predict their future in a new political environment. However, it is the author's belief that SACU is too much associated with apartheid South Africa and dominated by decision-making in Pretoria to provide a suitable model for future co-operation. In addition, South Africa has in the past been willing to support SACU through a system of generous compensatory payments to the participant states so as to develop closer links with southern Africa. This not only lessened the threat of sanctions but also generated an export market for the sale of South Africa's manufactured goods (Davies, 1990a). With apartheid and the sanctions threat removed, South Africa will feel less inclined to subsidize such countries when confronted with the need to devote increasing amounts of capital on its neglected black population. More importantly, however, South Africa could certainly not afford to replicate for the whole region the generous compensatory payments at present received by the BLS states (Botswana, Lesotho and Swaziland) and Namibia (Gibb, 1993). Similarly, the PTA is unlikely to provide a regional framework suitable for the needs of a post-apartheid southern Africa. The PTA lacks both geographical and structural coherence. With membership stretching from Lesotho in the south to the Horn of Africa, the PTA lacks a regional identity and cohesion. It is also hampered by a tremendous diversity in the socio-political backgrounds of its member states.

SADC has the benefit of being an 'indigenous' response to the need for regional cooperation and has the added political advantage of having its headquarters located outside South Africa. The present chapter therefore focuses exclusively on SADC(C) and the relevance of the European approach to integration. Before proceeding to examine the nature and purpose of SADC, it is first necessary to explore the approach SADCC adopted to regional integration. SADCC was the forerunner to SADC and built up a reputation for success which, although disputable, was widely recognised (Hanlon, 1989; Weimer, 1991). The transformation to SADC was based, in part, upon the problems and opportunities encountered by SADCC in its endeavour to promote co-operation and integration in the 1980s.

SADCC'S APPROACH TO INTEGRATION

SADCC started off its life by explicitly rejecting economic integration based on the European experience of trade liberalization and free-market principles.

Traditional economic integration theory or customs-union theory is based upon a neo-classical economic perspective that supports the free market, exploits economies of scale and promotes a more competitive environment. Such agreements are commonly based on an elimination of discriminatory measures between member-state economies. Balassa (1961) interprets economic integration as an evolutionary process comprised of a number of successive stages with each stage involving more complex and higher levels of integration. In theory, regional economic integration creates its own momentum with participating states eager to contemplate higher forms of economic integration so as to realize the benefits from enhanced trade liberalization and even greater economies of scale with respect to industrialization (Wise and Gibb, 1992). Although the progression from free trade area to economic and political union is widely recognized (Figure 11.2), it should be stressed that regional economic integration is a complex phenomenon which does not lend itself easily to simple theoretical formulations.

The original SADCC blueprint, as outlined in the 1980 Lusaka Declaration, rejected this model of economic integration based upon trade liberalization and stressed that SADCC was not a 'common market'. This rejection of common-market principles was consistently adhered to. In 1988, SADCC's chief executive, Dr Simba Makoni, stated:

> Our approach to trade in the region is not based on orthodox trade liberalization strategies . . . the reduction or even elimination of tariffs and other barriers to trade does not always yield increased trade, in the absence of tradeable goods' (quoted in Hanlon, 1989, p. 65).

	Removal of internal quota & tariff	Common external customs tariff	Free movements of land, labour, capital and services	Harmonization of economic policies and development of supranational institutions	Unification of and political and powerful super national institutions
Sectoral Cooperation	●				
Free Trade Association	●				
Customs Union	●	●			
Common Market	●	●	●		
Economic Union	●	●	●	●	
Political Union	●	●	●	●	●

Figure 11.2. Degrees of Regional Integration

The 1992 SADCC 'Theme Document' (SADCC, 1992) again rejected the traditional customs union common market approach, stating: 'A laissez-faire approach to regional integration, in a region with gross disparities, would be inappropriate, and tend to entrench existing inequalities and imbalances.'

In order to promote equitable regional development, SADCC deliberately avoided competition and 'unnecessary' duplication and pursued a policy of 'balanced trade' as opposed to free trade. Industrial production was to be managed, planned and co-ordinated as opposed to being exposed to the pressures of the free market. The cornerstone of this approach to integration was 'project co-ordination'. Emphasis was placed upon the need to promote projects in the spheres of production and infrastructure. In so doing, project co-ordination would encourage co-operation in regional development projects as 'a step or catalyst to integration' (SADCC, 1989). The SADCC approach to regional economic integration did not therefore fit anywhere into the traditional integration theory. Lee (1989) observed that SADCC was: 'the first serious attempt by a group of third world nations to reject totally traditional customs union or integration theory, or modifications thereof, as solutions to their regional problems.'

However, as the 1992 SADCC Theme Document made explicitly clear, project co-ordination failed to generate a movement towards closer regional co-operation or integration in either political or economic terms. The level of intra-SADCC trade actually fell during the mid-1980s (Table 11.1), and political co-operation remained at a minimal level. SADCC observed: 'The level of integration which currently exists among the ten SADCC Member States . . . remains extremely modest.' . . . 'Regional project coordination . . . cannot . . . be any longer the major basis of SADCC strategy. This is because project coordination has been recognised as having only a limited impact in promoting deeper or wider cooperation and integration.' (SADCC, 1992, p. 22) SADCC clearly recognized the need for integration within southern Africa: '. . . closer economic cooperation and integration have become no longer merely desirable, but imperative for growth, development and indeed survival' (p. 17).

Table 11.1. Intra-SADCC Trade

	SDR m	% of Total Trade
1981	548	4.7
1982	536	4.7
1983	495	4.5
1984	512	4.5
1985	417	3.8
1986	384	4.2

Source: SADCC, 1992.
SDR = Special drawing rights.

At the same time, however, SADCC recognized the failure of its project co-ordination strategy to promote integration and started to explore alternative integrative models (Kongwa, 1991). The SADCC secretariat identified what they referred to as a 'developmental integrative approach' as offering an attractive strategy for southern Africa. This approach appears to be a modified version of the 'project coordination' strategy but with greater emphasis on political co-operation at an early stage of the integration process. While the 'developmental integrative approach' continues to reject the *laissez-faire* principle on the grounds that it is an unsuitable mechanism for promoting integration among developing countries, the policies advocated bear a remarkable resemblance to some of those contained in the Treaty of Rome. It is interesting to compare the policy proposals of the 1992 SADCC Theme Document with the 1957 Rome Treaty:

SADCC Theme Document, Maputo, 1992

1. 'The freer movement of capital, goods and labour and people within the region.' (p. 1)
2. 'Ways of achieving a greater coordination of external tariff policy.' (p. 27)
3. 'Freedom of movement, residence and employment throughout the region.' (p. 29)
4. 'Trade liberalisation to be complemented by compensatory and corrective measures, orientated particularly towards the least developed countries.' (SADCC, 1992) (p. 22)

The Treaty of Rome, 1957

1. 'The abolition, as between Member States, of obstacles to freedom of movement for persons, services and capital goods.' (Article 3)
2. 'The establishment of a common customs tariff and a common commercial policy towards third countries.' (Article 3)
3. '. . . restrictions on the freedom of establishment of nationals of a Member State in the territory of another Member State shall be progressively abolished.' (Article 52)
4. 'Member States agree upon the necessity to promote improvement of the living and working conditions of labour so as to permit the equalization of such conditions in an upward direction.' (Article 117)
(Treaty of Rome, 1967)

In the early 1990s, although continuing to reject traditional integration theory based upon trade liberalization, SADCC started to move towards accepting, at least in principle, some elements of the free trade and liberalization argument. This reorientation was, in part, brought about by the policy initiatives of many of its member states, where the management of economic affairs was being slowly transformed to allow more market-orientated policies to prevail at the cost of state intervention.

The developmental integration approach encouraged the co-ordination of macroeconomic policies among member states at an early stage of development integration. Like the EC's '1992' programme, the SADCC Theme Document highlighted the advantages to be derived from a single regional market which would provide enhanced economies of scale, enabling a major review of investment strategies and priorities. In the political field, SADCC recognized that in the final analysis regional economic integration depends on the degree of political will and commitment of member states. Inevitably, as in all forms of integration, this would entail the transfer of sovereign powers to a regional supranational institution. Although SADCC did not identify any particular area of joint authority, it noted that:

> . . . regional bodies accountable to the central regional decision-making authorities will need to be established in each of the major sectors . . . Integration is sometimes seen as a process in which individual states lose sovereignty . . . however, this should more appropriately be seen as a change in the locus of exercising sovereignty, rather than a loss of sovereignty (SADCC, 1992, p. 33).

The need to pool sovereignty and share resources and productive capacities is at the very heart of any integrative process. SADCC's limited ability to promote deeper economic co-operation and integration throughout southern Africa highlighted a need to reform the organization.

Throughout its existence, SADCC solidarity has been built on the liberation struggles and a common apartheid threat. However, South Africa's abolition of statutory apartheid threatened this solidarity and SADCC's very *raison d'etre*. The wide-ranging policy proposals formulated by the SADCC secretariat in 1992 were, in part, a response to that threat. They also reflected a desire to have in place a working and operationally recognized structure before South Africa became 'the eleventh member state'. SADCC claimed that: 'Having in place a functioning programme and effective institutions of integration could well be decisive in determining whether it will be South Africa that joins SADCC, or SADCC that joins South Africa' (SADCC, 1992, p. 20).

However, the desire to promote deeper economic co-operation and integration arose also from a renewed realization of the benefits to be derived from restructuring member states' economies on a regional scale.

SADC'S APPROACH TO INTEGRATION

It is too early to examine in detail the integrative strategy of the new SADC. However, it is possible to identify a few basic principles outlined in the Treaty (SADC, 1992b) and the 'Declaration' (SADC, 1992a) signed by the Southern African Heads of State. The Declaration recognizes clearly the merits of closer regional cooperation:

Integration is fast becoming a global trend . . . movements towards stronger regional blocs will transform the world, both economically and politically. Firms within these economic blocs will benefit from economies of scale provided by large markets, to become competitive both internally and internationally (SADC, 1992a, p. 3).

SADC also recognized that, thus far, progress towards regional integration had been modest and that the region has failed to mobilize its own resources to the fullest extent possible. In order to deepen regional integration, SADC recognized the need both to enhance the political commitment of member states and to establish effective institutions and mechanisms to mobilize the region's own resources. Translating this into policy terms, SADC plans, among other things, to:

• (a) harmonize political and social-economic policies and plans of member states
• (b) develop policies aimed at the progressive elimination of obstacles to the free movement of capital and labour, goods and services
• (c) promote the coordination and harmonization of the international relations of member states (SADC, 1992b, Article 5)

The SADC Treaty therefore established a framework of co-operation that prioritizes: deeper economic co-operation and integration; common economic and political systems; and a common foreign policy. The new SADC promises to be more powerful than the old SADCC, possessing limited supranational powers in both economic and political fields.

The extent to which genuine supranational institutions have been established by the SADC Treaty is unclear. The two most powerful institutions of SADC are 'The Summit' and 'The Council of Ministers'. Both of these institutions are intergovernmental in character comprising one minister, or the heads of government/state in the case of the Summit, from each member state. The Treaty refers to the Summit as 'the supreme policy-making institution of SADC' (SADC, 1992b, Article 10.1). However, the degree to which these institutions represent an effective transfer of sovereignty depends on the voting systems adopted. The SADC Treaty is less than clear on the voting system, stating that: 'Unless otherwise provided in this Treaty, the decisions of the Summit shall be made by consensus and shall be binding' and 'Decisions of the Council shall be by consensus' (SADC, 1992b, pp. 11, 13).

According to the *Oxford English Dictionary*, 'consensus' refers to ' a general agreement in opinion'. It is therefore unclear whether decisions will be taken under majority voting or qualified majority voting and whether the right of veto exists on issues of vital national importance. Until the operating conditions of these institutions become established, it remains too early to evaluate their supranational powers. However, as the Declaration makes explicitly clear, SADC's intention is to have supranational powers:

Integration does imply that some decisions which were previously taken by individual states are taken regionally . . . Regional decision-making also implies
elements of change in the locus and content of exercising sovereignty (SADC, 1992a,
p. 10).

The 'developmental integration strategy' adopted by SADC is apparently
dependent on the establishment of mechanisms 'capable of achieving a high level
of political cooperation and conscious intervention' (SADCC, 1992). For
example, the establishment and co-ordination of a macroeconomic policy, even if
restricted to a few specific areas, would be a significant integrative mechanism
demanding a substantial pooling of sovereignty among member states. SADC
therefore requires the establishment of strong regional institutions that are
accountable not to individual member states but to central regional decision-
making authorities.

IN CONCLUSION

While SADC has accepted at least some of the *laissez-faire* and trade liberalization strategies underpinning the customs-union type approach to integration,
it has rejected the traditional model of integration based upon free-market forces
projecting a group of countries from free-trade area to economic and political
union (see Figure 11.2). SADC is promoting policies that are usually considered
as occurring only at a fairly late stage in the conventional trade integration
approach. It is here that the European experience may be of value. As already
stated, the purpose of this chapter is neither to promote a Eurocentric view of
integration nor to transplant the traditional customs-union type model to the
southern African region. However, SADC could and should learn from the
European experience of integration. In particular, the advantages of adopting the
neo-functionalist spillover strategy would appear to be particularly appropriate.
Jean Monnet and Robert Schumann recognized the impossibility of creating
European unity in one mighty constitutional stroke that enforced supranational
and federal structures overnight. The policy proposals of SADC, which create a
macroeconomic policy, promote political integration and create supranational
institutions even before an effectively functioning trade area has been established,
may well therefore turn out to be politically impracticable. In effect, the policies
now being pursued by SADC are along the lines of the European federalist
strategies pursued in the immediate aftermath of World War Two. Consequently,
it may be more appropriate to adopt policies of a more limited nature that
establish integrative bodies that have specifically defined and limited supranational powers over a particular function such as the ECSC. Agriculture, tourism,
energy, transport, water and the environment are all areas within southern Africa
where this approach to integration appears feasible. If the political will exists,
then spillover will quickly lead to activity in other areas of the economy and, in

the long term, to greater integration within the region. SADC should therefore consider adopting the neo-functionalist approach to regional integration. In 1950, commenting on the prospects for European unity, Schumann (1990) stated that 'Europe will not be made at once, or according to a single plan. It will be built through concrete achievements which will first create a de facto identity'. The same is true for southern Africa.

NOTE

1. The 'Constellation of States of Southern Africa' (CONSAS) was first proposed by President P.W. Botha at the Carlton Conference in Johannesburg, 1979. CONSAS was an ill-defined concept that promoted an ill-defined form of regionalism, proposing a 'grouping' rather than a formal organization. It tried to promote a form of regionalism that involved various levels of integration for different member states (to use the terminology common in the European Community, a 'variable geometry' southern Africa). Within two years the CONSAS concept had drifted into obscurity and, in common with most other South African attempts to promote regionalism, overestimated the integrative power of economic ties to override political and ideological differences.

REFERENCES

Balassa, B. (1961) *The Theory of Economic Integration*, Irwin, Homewood, Illinois.
Blumenfeld, J. (1991) *Economic Interdependence in Southern Africa: from Conflict to Cooperation*, Pinter, London.
Cecchini, P. (1988) *1992: the Benefits of a Single Market*, Wildwood House, Aldershot.
Davies, R. (1990a) 'Post-apartheid scenarios for the Southern African region', *Transformation*, 11, 12–39.
Davies, R. (1990b) *Key Issues in Reconstructing South–Southern African Relations after Apartheid*, working paper series, Centre for Southern African Studies, University of the Western Cape, Bellville.
Davies, R. (1991) 'Southern Africa into the 1990s and beyond', paper presented to a conference on 'Current and Future Prospects for the Political Economy of Southern Africa', Broederstroom, 15–19 May.
Davenport, M. (1992) 'Africa and the unimportance of being preferred', *Journal of Common Market Studies*, 30(2), 233–51.
du Pisani, A. (1991) 'Ventures into the interior: continuity and change in South Africa's regional policy 1948–1991', paper presented to the conference 'Southern Africa into the 1990s and Beyond', Magaliesburg, 15–19 April.
George, S. (1985) *Politics and Policy in the European Community*, Clarendon Press, Oxford.
Gibb, R.A. (1992) 'Regional integration must evolve one step at a time', *Business Day*, 13 August.
Gibb, R.A. (1993) 'A common market for post-apartheid Southern Africa: prospects and problems', *South African Geographical Journal*, 75(1), 28–35.
Green, R.H. (1981) 'First steps towards economic liberation' in A.J. Nsekela (ed.), *Southern Africa: Toward Economic Liberation*, Rex Collings, London.

Hanlon, J. (1989) *SADCC in the 1990s: Development on the Front Line*, special report no. 1158, The Economist Intelligence Unit, London.

Holland, M. (1983) 'A Rejoinder: the European Community and regional integration—misplaced analogy', *Politikon: South African Journal of Political Science*, 10(2), 38–50.

Iconescu, C. (1972) *The Politics of European Integration*, Macmillan, London.

Kongwa, S. (1991) 'SADCC: creating a new vision for the future', *Africa Institute Bulletin*, 16(10), 1.

Langhammer, R.J. (1992) 'The developing countries and regionalism', *Journal of Common Market Studies*, 30(2), 211–31.

Lee, C. (1989) Options for Regional Cooperation and Development in Southern Africa, unpublished Ph.D. thesis, University of Pittsburgh.

Leistner, G.M.E. (1992) 'Post-apartheid South Africa's economic ties with neighbouring countries', *Development Southern Africa*, 9(2), 169–87.

Lindberg, L. (1963) *The Political Dynamics of European Integration*, Oxford University Press, Oxford.

Maasdorp, G. (1986) 'The Southern African nexus: dependence or interdependence?', *Indicator South Africa: Economic Monitor*, 4:5–19.

Maasdorp, G. (1989) 'Economic relations in Southern Africa—changes ahead?', paper presented to the conference on 'South and Southern Africa in the 21st century', Maputo, December.

Maasdorp, G. (1992) 'Economic prospects for South Africa in Southern Africa', *South Africa International*, 22(3), 121–7.

Matthews, J. (1984) 'Economic integration in Southern Africa: progress or decline?', *South African Journal of Economics*, 52(3), 256–65.

Mitrany, D. (1933) *The Progress of International Government*, Allen and Unwin, London.

Mitrany, D. (1965) 'The prospects of integration: federal or functional', *Journal of Common Market Studies*, 4(1), 119–50.

Mitrany, D. (1966) *A Working Peace System*, Royal Institute of International Affairs, London.

Mitrany, D. (1975) *The Functional Theory of Politics*, Robertson, London.

Perry, K. (1984) *Britain and the European Community*, Heinemann, London.

SADC (1992a) *Towards the Southern African Development Community: a Declaration*, SADC, Gaborone.

SADC (1992b) *Treaty of the Southern African Development Community*, SADC, Gaborone.

SADCC (1981) *SADCC2–Maputo*, SADCC Liaison Committee, SADCC, Gabarone.

SADCC (1989) *SADCC Annual Progress Report 1989–90*, SADCC, Gaborone.

SADCC (1992) *SADCC Theme Document*, Maputo conference, 29–31 January, SADCC, Gaborone.

Scheepers, C. (1979) 'The possible role of a customs-union type model in promoting close economic ties in Southern Africa', *Finance and Trade Review*, 13(4), 82–9.

Schumann, R. (1990) *The Schumann Declaration*, reprinted in *Europe: a Fresh Start*, Europe Publication 3/1990, office for official publications of the EC, Luxembourg.

Tjonneland, E.N. (1989) 'South Africa's regional policies in the late post-apartheid period' in B. Oden and H. Othman (eds), *Regional Co-operation in Southern Africa: a Post-apartheid Perspective*, Scandinavian Institute of African Studies, seminar proceedings, no. 22, Uppsala.

Treaty of Rome (1967) *Treaty of Rome Establishing the European Economic Community*, HMSO, London.

Vale, P. (1982) 'Prospects for transplanting European models of regional integration to Southern Africa', *Politikon: South African Journal of Political Science*, 9(2), 31–42.

Weimer, B. (1991) 'The Southern African Development Coordination Conference (SADCC): past and future', *Africa Insight*, **21**(2), 78–88.

Wise, M. (1994) 'The European Community' in R.A. Gibb and W.Z. Michalek (eds), *Continental Trading Blocs: the Growth of Regionalism in the World Economy*, John Wiley, Chichester, 75–110.

Wise, M. and Gibb, R.A. (1992) *Single Market to Social Europe: The European Community in the 1990s*, Longman, London.

Index